# One Hundred Nineteen Stata Tips

## Third Edition

T0186305

# One Hundred Nineteen Stata Tips

## Third Edition

NICHOLAS J. COX, Editor
*Durham University*
*Department of Geography*

H. JOSEPH NEWTON, Editor
*Texas A & M University*
*Department of Statistics*

A Stata Press Publication
StataCorp LP
College Station, Texas

Published by Stata Press, 4905 Lakeway Drive, College Station, Texas 77845
Typeset in LATEX $2_\varepsilon$
Printed in the United States of America

10 9 8 7 6 5 4 3 2 1

ISBN-10: 1-59718-143-9
ISBN-13: 978-1-59718-143-3

Library of Congress Control Number: 2014936591

# Contents

# Subject table of contents

# General

# Data management

# Graphics

# Programming

# Statistics

# Editors' preface

The book you are reading reprints 119 Stata Tips from the *Stata Journal*, with thanks to their original authors. We, the *Journal* editors, began publishing tips in 2003, beginning with volume 3, number 4. It pleases us now to reprint them in this book.

The *Stata Journal* publishes substantive and peer-reviewed articles ranging from reports of original work to tutorials on statistical methods and models implemented in Stata, and indeed on Stata itself. Other features include regular columns such as "Speaking Stata", book reviews, and announcements.

We are pleased by the external recognition that the *Journal* has achieved. The *Stata Journal* is indexed and abstracted by CompuMath Citation Index, Current Contents/Social and Behavioral Sciences, RePEc: Research Papers in Economics, Science Citation Index Expanded (also known as SciSearch), Scopus, and Social Sciences Citation Index.

But back to the Tips. There was little need for tips in the early days. Stata 1.0 was released in 1985. The original program had 44 commands, and its documentation totaled 175 pages. Stata 13, on the other hand, has more than 1,000 commands—including an embedded matrix language called Mata—and Stata's official documentation now totals more than 11,000 pages. Beyond that, the user community has added several hundred more commands.

The pluses and the minuses of this growth are evident. As Stata expands, it is increasingly likely that users' needs can be met by available code. But at the same time, learning how to use Stata and even learning what is available become larger and larger tasks.

The Tips are intended to help. The ground rules for Stata Tips, as found in the original 2003 statement, are laid out as the next item in this book. We have violated one original rule in the letter, if not the spirit: Some Stata Tips have been as long as six pages. However, the intention of producing concise tips that are easy to pick up remains as it was.

The Tips grew from many discussions and postings on Statalist, at Users Group meetings, and elsewhere, which underscores a simple fact: Stata is now so big that it is easy to miss even simple features that can streamline and enhance your sessions with Stata. This applies not just to new users, who understandably may quake nervously before the manual mountain, but also to longtime users, who too are faced with a mass of new features in every release.

Tips have come from Stata users as well as from StataCorp employees. Many discuss new features of Stata, or features not documented fully or even at all. We hope that you enjoy the Stata Tips reprinted here and can share them with your fellow Stata users. If you have tips that you would like to write, or comments on the kinds of tips that are helpful, do get in touch with us, as we are eager to continue the series.

Nicholas J. Cox, Editor
H. Joseph Newton, Editor
April 2014

The Stata Journal (2003)
**3**, Number 4, p. 328

# Introducing Stata tips

As promised in our editorial in *Stata Journal* 3(2), 105–108 (2003), the *Stata Journal* is hereby starting a regular column of tips. Stata tips will be a series of concise notes about Stata commands, features, or tricks that you may not yet have encountered.

The examples in this issue should indicate the kinds of tips we will publish. What we most hope for is that readers are left feeling, "I wish I had known that earlier!" Beyond that, here are some more precise guidelines:

**Content** A tip will draw attention to useful details in Stata or in the use of Stata. We are especially keen to publish tips of practical value to a wide range of users. A tip could concern statistics, data management, graphics, or any other use of Stata. It may include advice on the user interface or about interacting with the operating system. Tips may explain pitfalls (do not do this) as well as positive features (do use this). Tips will not include plugs for user-written programs, however smart or useful.

**Length** Tips must be brief. A tip will take up at most three printed pages. Often a code example will explain just as much as a verbal discussion.

**Authorship** We welcome submissions of tips from readers. We also welcome suggestions of tips or of kinds of tips you would like to see, even if you do not feel that you are the person to write them. Naturally, we also welcome feedback on what has been published. An email to *editors@stata-journal.com* will reach us both.

H. Joseph Newton, Editor
Texas A&M University
jnewton@stat.tamu.edu


Nicholas J. Cox, Editor
University of Durham
n.j.cox@durham.ac.uk

# Stata tip 1: The eform() option of regress

Roger Newson, King's College London, UK
roger.newson@kcl.ac.uk

Did you know about the `eform()` option of `regress`? It is very useful for calculating confidence intervals for geometric means and their ratios. These are frequently used with skewed $Y$-variables, such as house prices and serum viral loads in HIV patients, as approximations for medians and their ratios. In Stata, I usually do this by using the `regress` command on the logs of the $Y$-values, with the `eform()` and `noconstant` options. For instance, in the `auto` dataset, we might compare prices between non-US and US cars as follows:

```
. sysuse auto, clear
(1978 Automobile Data)
. generate logprice = log(price)
. generate byte baseline = 1
. regress logprice foreign baseline, noconstant eform(GM/Ratio) robust
Regression with robust standard errors          Number of obs =      74
                                                 F(  2,    72) =18043.56
                                                 Prob > F      =  0.0000
                                                 R-squared     =  0.9980
                                                 Root MSE      = .39332
```

| logprice | GM/Ratio | Robust Std. Err. | t | P>\|t\| | [95% Conf. Interval] | |
|---|---|---|---|---|---|---|
| foreign | 1.07697 | .103165 | 0.77 | 0.441 | .8897576 | 1.303573 |
| baseline | 5533.565 | 310.8747 | 153.41 | 0.000 | 4947.289 | 6189.316 |

We see from the `baseline` parameter that US-made cars had a geometric mean price of 5534 dollars (95% CI from 4947 to 6189 dollars), and we see from the `foreign` parameter that non-US cars were 108% as expensive (95% CI, 89% to 130% as expensive). An important point is that, if you want to see the baseline geometric mean, then you must define the constant variable, here `baseline`, and enter it into the model with the `noconstant` option. Stata usually suppresses the display of the intercept when we specify the `eform()` option, and this trick will fool Stata into thinking that there is no intercept for it to hide. The same trick can be used with `logit` using the `or` option, if you want to see the baseline odds as well as the odds ratios.

My nonstatistical colleagues understand regression models for log-transformed data a lot better this way than any other way. Continuous $X$-variables can also be included, in which case the parameter for each $X$-variable is a ratio of $Y$-values per unit change in $X$, assuming an exponential relationship—or assuming a power relationship, if $X$ is itself log-transformed.

The Stata Journal (2003)
**3**, Number 4, pp. 446–447

# Stata tip 2: Building with floors and ceilings

Nicholas J. Cox, University of Durham, UK
n.j.cox@durham.ac.uk

Did you know about the `floor()` and `ceil()` functions added in Stata 8?

Suppose that you want to round down in multiples of some fixed number. For concreteness, say, you want to round `mpg` in the auto data in multiples of 5 so that any values 10–14 get rounded to 10, any values 15–19 to 15, etc. `mpg` is simple, in that only integer values occur; in many other cases, we clearly have fractional parts to think about as well.

Here is an easy solution: `5 * floor(mpg/5)`. `floor()` always rounds down to the integer less than or equal to its argument. The name floor is due to Iverson (1962), the principal architect of APL, who also suggested the expressive $\lfloor x \rfloor$ notation. For further discussion, see Knuth (1997, 39) or Graham, Knuth, and Patashnik (1994, chapter 3).

As it happens, `5 * int(mpg/5)` gives exactly the same result for `mpg` in the auto data, but in general, whenever variables may be negative as well as positive, *interval* * `floor(`*expression/interval*`)` gives a more consistent classification.

Let us compare this briefly with other possible solutions. `round(mpg, 5)` is different, as this rounds to the nearest multiple of 5, which could be either rounding up or rounding down. `round(mpg - 2.5, 5)` should be fine but is also a little too much like a dodge.

With `recode()`, you need two dodges, say, `-recode(-mpg,-40,-35,-30,-25,-20,-15,-10)`. Note all the negative signs; negating and then negating to reverse it are necessary because `recode()` uses its numeric arguments as upper limits; i.e., it rounds up.

`egen, cut()` offers another solution with option call `at(10(5)45)`. Being able to specify a *numlist* is nice, as compared with spelling out a comma-separated list, but you *must* also add a limit, here 45, which will not be used; otherwise, with `at(10(5)40)`, your highest class will be missing.

Yutaka Aoki also suggested to me `mpg - mod(mpg,5)`, which follows immediately once you see that rounding down amounts to subtracting the appropriate remainder. `mod(,)`, however, does not offer a correspondingly neat way of rounding up.

The `floor` solution grows on one, and it has the merit that you do not need to spell out all the possible end values, with the risk of forgetting or mistyping some. Conversely, `recode()` and `egen, cut()` are not restricted to rounding in equal intervals and remain useful for more complicated problems.

Without recapitulating the whole argument insofar as it applies to rounding up, `floor()`'s sibling `ceil()` (short for ceiling) gives a nice way of rounding up in equal intervals and is easier to work with than expressions based on `int()`.

# References

Graham, R. L., D. E. Knuth, and O. Patashnik. 1994. *Concrete Mathematics: A Foundation for Computer Science*. Reading, MA: Addison–Wesley.

Iverson, K. E. 1962. *A Programming Language*. New York: Wiley.

Knuth, D. E. 1997. *The Art of Computer Programming, Volume 1: Fundamental Algorithms*. Reading, MA: Addison–Wesley.

The Stata Journal (2003)
**3**, Number 4, p. 448

# Stata tip 3: How to be assertive[1]

William Gould, StataCorp
wgould@stata.com

assert verifies the truth of a claim:

```
. assert sex=="m" | sex=="f"
. assert age>=18 & age<=65
22 contradictions in 2740 observations
assertion is false
r(9);
```

The best feature of assert is that, when the claim is false, it stops do-files and ado-files:

```
. do my_data_prep
. use basedata, clear
. assert age>=18 & age<=64
22 contradictions in 2740 observations
assertion is false
r(9);
end of do-file
r(9);
```

assert has two main uses:

1. It checks that claims made to you and suppositions you have made about the data you are about to process are true:

   ```
   . assert exp==. if age<18
   . assert exp<. if age>=18
   ```

2. It tests that, when you write complicated code, the code produces what you expect:

   ```
   . sort group
   . by group: gen avg = sum(hours)/sum(hours<.)
   . by group: assert avg!=. if _n==_N
   . by group: gen relative = hours/avg[_N]
   ```

assert is especially useful following merge:

```
. merge m:1 id using demog
. assert _merge==3
. drop _merge
```

---

1. This tip was updated to reflect the new merge syntax.—Ed.

The Stata Journal (2004)
**4**, Number 1, p. 93

# Stata tip 4: Using display as an online calculator

Philip Ryan, University of Adelaide
philip.ryan@adelaide.edu.au

Do you use Stata for your data management, graphics, and statistical analysis but switch to a separate device for quick calculations? If so, you might consider the advantages of using Stata's built-in `display` command:

1. It is always at hand on your computer.

2. As with all Stata calculations, double precision is used.

3. You can specify the format of results.

4. It uses and reinforces your grasp of Stata's full set of built-in functions.

5. You can keep an audit trail of results and the operations that produced those results, as part of a log file. You can also add extra comments to the output.

6. Editing of complex expressions is easy, without having to re-enter lengthy expressions after a typo.

7. You can copy and paste results elsewhere whenever your platform supports that.

8. It is available via the menu interface (select **Data—Other utilities—Hand calculator**).

9. It can be abbreviated to `di`.

To be fair, there are some disadvantages, such as its lack of support for Reverse Polish Notation or complex number arithmetic, but in total, `display` provides you with a powerful but easy-to-use calculator.

```
. di _pi
3.1415927
. di %12.10f _pi
3.1415926536
. * probability of 2 heads in 6 tosses of a fair coin
. di comb(6,2) * 0.5^2 * 0.5^4
.234375
. di "chi-square (1 df) cutting off 5% in upper tail is " invchi2tail(1, .05)
chi-square (1 df) cutting off 5% in upper tail is 3.8414588
. * Euler-Mascheroni gamma
. di %12.10f -digamma(1)
0.5772156649
```

The Stata Journal (2004)
**4**, Number 1, p. 94

# Stata tip 5: Ensuring programs preserve dataset sort order

Roger Newson, King's College London, UK
roger.newson@kcl.ac.uk

Did you know about `sortpreserve`? If you are writing a Stata program that temporarily changes the order of the data and you want the data to be sorted in its original order at the end of execution, you can save a bit of programming by including `sortpreserve` on your `program` statement. If your program is called `myprogram`, you can start it with

```
program myprogram, sortpreserve
```

If you do this, you can change the order of observations in the dataset in `myprogram`, and Stata will automatically sort it in its original order at the end of execution. Stata does this by creating a temporary variable whose name is stored in a macro named `_sortindex`, which is discussed in the manuals under [P] **sortpreserve**. (Note, however, that there is a typo in the manual; the underscore in `_sortindex` is missing.[1]) The temporary variable '`_sortindex`' contains the original sort order of the data, and the dataset is sorted automatically by '`_sortindex`' at the end of the program's execution.

If you know about temporary variables, you might think that `sortpreserve` is unnecessary because you can always include two lines at the beginning, such as

```
tempvar order
generate long `order' = _n
```

and a single line at the end such as

```
sort `order'
```

and do the job of `sortpreserve` in 3 lines. However, `sortpreserve` does more than that. It restores the result of the macro extended function `sortedby` to the value that it would have had before your program executed. (See [P] **macro** for a description of `sortedby`.) Also, it restores the "Sorted by:" variable list reported by the `describe` command to the variable list that would have been reported before your program executed. For example, in the `auto` dataset shipped with official Stata, the output of `describe` ends with the message

```
Sorted by:  foreign
```

This will not be changed if you execute a program defined with `sortpreserve`.

---

1. This has been fixed in Stata 10 and later manuals.

The Stata Journal (2004)
4, Number 1, pp. 95–96

# Stata tip 6: Inserting awkward characters in the plot[1]

Nicholas J. Cox, University of Durham, UK
n.j.cox@durham.ac.uk

Did you know about the function `char()`? `char(`$n$`)` returns the character corresponding to ASCII code $n$ for $1 \leq n \leq 255$. There are several numbering schemes for so-called ASCII characters. Stata uses the ANSI scheme; a web search for "ANSI character set" will produce tables showing available characters. This may sound like an arcane programmer's tool, but it offers a way to use awkward text characters—either those not available through your keyboard or those otherwise problematic in Stata. A key proviso, however, is that you must have such characters available in the font that you intend to use. Fonts available tend to vary not only with platform but even down to what is installed on your own system. Some good fonts for graphics, in particular, are Arial and Times New Roman.

Let us see how this works by considering the problem of inserting awkward characters in your Stata graphs, say as part of some plot or axis title. Some examples of possibly useful characters are

| | | |
|---|---:|---:|
| `char(133)` | ellipsis | . . . |
| `char(134)` | dagger | † |
| `char(135)` | double dagger | ‡ |
| `char(169)` | copyright | © |
| `char(176)` | degree symbol | ° |
| `char(177)` | plus or minus | ± |
| `char(178)` | superscript 2 | $^2$ |
| `char(179)` | superscript 3 | $^3$ |
| `char(181)` | micro symbol | $\mu$ |
| `char(188)` | one-fourth | $\frac{1}{4}$ |
| `char(189)` | one-half | $\frac{1}{2}$ |
| `char(190)` | three-fourths | $\frac{3}{4}$ |
| `char(215)` | multiply | × |

There are many others that might be useful to you, including a large selection of accented letters of the alphabet. You can use such characters indirectly or directly. The indirect way is to place such characters in a local macro and then to refer to that macro within the same program or do-file.

For example, I use data on river discharge, for which the standard units are cubic meters per second. I can get the cube power in an axis title like this:

---

1. Stata 11 introduced a much easier way to add symbols and Greek letters to graph text. See [G-4] *text* or `help graph text`.—Ed.

```
. local cube = char(179)
. scatter whatever, xtitle("discharge, m`cube´/s")
```

Or, I have used Hanning, a binomial filter of length 3:

```
. local half = char(189)
. local quarter = char(188)
. twoway connected whatever,
>         title("Smoothing with weights `quarter´:`half´:`quarter´")
```

The direct way is to get a macro evaluation on the fly. You can write '=char(176)' and, in one step, get the degree symbol (for temperatures or compass bearings). This feature was introduced in Stata 7 but not documented until Stata 8. See [P] **macro**.

char() has many other uses besides graphs. Suppose that a string variable contains fields separated by tabs. For example, insheet leaves tabs unchanged. Knowing that a tab is char(9), we can

```
. split data, p(`=char(9)´) destring
```

Note that p(char(9)) would not work. The argument to the parse() option is taken literally, but the function is evaluated on the fly as part of macro substitution.

Note also that the SMCL directive {c #} may be used for some but not all of these purposes. See [P] **smcl**. Thus,

```
. scatter whatever, xtitle("discharge, m{c 179}/s")
```

would work, but using a SMCL directive would not work as desired with split.

The Stata Journal (2004)
4, Number 2, p. 220

# Stata tip 7: Copying and pasting under Windows[1]

Shannon Driver          Patrick Royston
StataCorp               MRC Clinical Trials Unit, London
sdriver@stata.com       patrick.royston@ctu.mrc.ac.uk

Windows users often copy, cut, and paste material between applications or between windows within applications. Here are two ways you can do this with Stata for Windows. We will describe one as a mouse-and-keyboard operation and the other as a menu-based operation. Experienced Windows users will know that these methods are, to a large extent, alternatives.

First, you can highlight some text in the Results window, copy it using the mouse (or keyboard), and then paste it into the Command window, the Do-file Editor, or anywhere else appropriate. This is a convenient way to transfer, for example, single values, lists, or sets of variable names from the screen for use in the next command. To copy text, place your mouse at the beginning of the desired text, drag to the end, thus highlighting the selected text, and press Ctrl-C. To paste text, click your mouse at the appropriate place and press Ctrl-V.

Suppose that a local macro 'macro' holds some text you wish to use. Then type

```
. display "`macro'"
```

and copy and paste the contents of 'macro' for editing in the Command window. Or, list in alphabetic order the names of variables not beginning with _I:

```
. ds _I*, not alpha
```

and then copy and paste the list into the Do-file Editor.

Second, suppose that you want to save a table constructed using tabstat in a form that makes it easy to convert into a table in MS Word. Stata has a **Copy Table** feature that you might find very useful. Make sure at the outset that you have set suitable options by clicking **Edit** in the menu bar and then **Table Copy Options**. In this case, removing all the vertical bars is advisable, so make sure **Remove all** is selected, and click **OK**. Now highlight the table in the Results window, and then click **Edit** and then **Copy Table**.

In MS Word, click **Edit** and then **Paste**. Highlight the pasted text and then click **Table** and then **Convert** and **Text to Table**. Specify **Tabs** under the **Separate text at** if it is not already selected. Click **OK** to create your table.

---

1. As of Stata 11, Stata now has two additional features to make copying easier, **Copy Table as HTML** and **Copy as Picture**.—Ed.

The Stata Journal (2004)
**4**, Number 2, pp. 221–222

# Stata tip 8: Splitting time-span records with categorical time-varying covariates

Ben Jann, ETH Zürich, Switzerland
jann@soz.gess.ethz.ch

In survival analysis, time-varying covariates are often handled by the method of episode splitting. The `stsplit` command does this procedure very well, especially in the case of continuous time-varying variables such as age or time in study. Quite often, however, we are interested in evaluating the effect of a change in some kind of categorical status or the occurrence of some secondary event. For example, we might be interested in the effect of the birth of a child on the risk of divorce or the effect of having completed further education on the chances of upward occupational mobility.

In such situations, the creation of splits might appear to be more complicated, and `stsplit` does not seem to be of much help, at least judging from the rather complicated examples provided with [ST] **stset** (*Final example: Stanford heart transplant data*) and [ST] **stsplit** (*Example 3: Explanatory variables that change with time*). Fortunately, the procedure is simpler than it appears.

Consider the Stanford heart transplant data used in examples for [ST] **stset** and [ST] **stsplit**:

```
. use http://www.stata-press.com/data/r8/stanford, clear
(Heart transplant data)

. list id transplant wait stime died if id==44 | id==16
```

|      | id | transp~t | wait | stime | died |
|------|----|----------|------|-------|------|
| 33.  | 44 | 0        | 0    | 40    | 1    |
| 34.  | 16 | 1        | 20   | 43    | 1    |

The goal here is to split the single time-span records into episodes before and after transplantation (e.g., to split case 16 at time 20). This can easily be achieved by splitting "at 0" "after `wait`", the time of transplantation. Note that, if no transplantation was carried out at all, `wait` should be recoded to a value larger than the observed maximum episode duration (maximum of `stime`) before the `stsplit` command is applied:

```
. replace wait = 10000 if wait == 0
(34 real changes made)

. stset stime, failure(died) id(id)
  (output omitted)

. stsplit posttran, after(wait) at(0)
(69 observations (episodes) created)

. replace posttran = posttran + 1
(172 real changes made)
```

```
. list id _t0 _t posttran if id == 44 | id == 16
```

|      | id | _t0 | _t | posttran |
|------|----|-----|-----|----------|
| 23.  | 16 | 0   | 20  | 0        |
| 24.  | 16 | 20  | 43  | 1        |
| 70.  | 44 | 0   | 40  | 0        |

It is now possible to evaluate the effect of transplantation on survival time using streg, for example, or to plot survivor functions with time-dependent group membership:

```
. sts graph, by(posttran)
         failure _d:  died
   analysis time _t:  stime
                id:  id
```

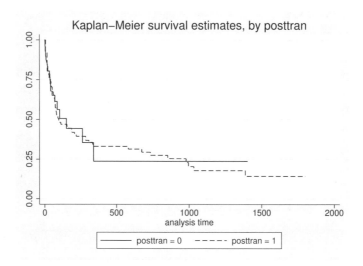

Note that the situation is even simpler if a Cox proportional hazards model is to be fitted. As explained in [ST] **stsplit**, the partial likelihood estimator takes only the times at which failures occur into account. Thus, in the context of Cox regression, the following code would do:

```
. use http://www.stata-press.com/data/r8/stanford, clear
. stset stime, failure(died) id(id)
. stsplit, at(failures)
. generate posttran = wait<_t & wait!=0
. stjoin
. stcox age posttran surgery year
```

The Stata Journal (2004)
**4**, Number 2, p. 223

# Stata tip 9: Following special sequences

Nicholas J. Cox
University of Durham, UK
n.j.cox@durham.ac.uk

Did you know about the special sequences stored as c-class values? [P] **creturn** documents various constant and current values, which may be seen by `creturn list` or which may be accessed once you know their individual names. For example, c(`filename`) stores the name of the file last specified with a `use` or `save` in the current session. However, various special sequences have been added in updates on 1 July 2003 and 15 December 2003 and so are not documented in the manuals. Here is the list:

- c(`alpha`) returns a string containing a space-separated list of the lowercase letters.

- c(`ALPHA`) returns a string containing a space-separated list of the uppercase letters.

- c(`Mons`) returns a string containing a space-separated list of month names abbreviated to three characters.

- c(`Months`) returns a string containing a space-separated list of month names.

- c(`Wdays`) returns a string containing a space-separated list of weekday names abbreviated to three characters.

- c(`Weekdays`) returns a string containing a space-separated list of weekday names.

Even the display of one of these lists can be useful. Note the local macro notation ' ' ensuring that the contents of the list are shown, not its name:

```
. display "`c(Months)'"
```

A common application of these lists is specifying variable or value labels. Suppose that a variable `month` included values 1 to 12. We might type

```
. tokenize `c(Months)'
. forvalues i = 1/12 {
  2. label def month `i' "``i''" , modify
  3. }
. label val month month
```

Finally, the SSC archive (see [R] **ssc**) is organized alphabetically using folders **a** through **z** and **_**. We could get a complete listing of what was available by

```
. foreach l in `c(alpha)' _ {
  2. ssc describe `l'
  3. }
```

# Stata tip 10: Fine control of axis title positions

Philip Ryan
University of Adelaide
philip.ryan@adelaide.edu.au

Nicholas Winter
Cornell University
nw53@cornell.edu

`ytitle()`, `xtitle()`, and other similar options specify the titles that appear on the axes of Stata graphs (see [G-3] *axis_title_options*). Usually, Stata's default settings produce titles with a satisfactory format and position relative to the axis. Sometimes, however, you will need finer control over position, especially if there is inadequate separation of the title and the numeric axis labels. This might happen, for example, with certain combinations of the font of the axis labels, the angle the labels make with the axis, the length of the labels, and the size of the graph region.

Although the options `ylabel()` and `xlabel()` have a suboption `labgap()` allowing user control of the gap between tick marks and labels (see [G-3] *axis_label_options*), the axis title options have no such suboption. The flexibility needed is provided by options controlling the textbox that surrounds the axis title (see [G-3] *textbox_options*). This box is invisible by default but can be displayed using the `box` suboption on the axis title option:

```
. graph twoway scatter price weight,
        ytitle("Price of Cars in {c S|}US", box)
        ylab(0(1000)15000, angle(horizontal) labsize(medium))
```

(Note the use of a SMCL directive to render the dollar sign; see [P] **smcl**, page 393.) We can manipulate the relative size of the height of the textbox or the margins around the text within the box to induce the appearance of a larger or smaller gap between the axis title and the axis labels. For a larger gap, we might try one of these solutions:

```
. graph twoway scatter price weight,
        ytitle("Price of Cars in {c S|}US", height(10))
        ylab(0(1000)15000, ang(hor) labsize(medium))
. graph twoway scatter price weight,
        ytitle("Price of Cars in {c S|}US", margin(0 10 0 0))
        ylab(0(1000)15000, ang(hor) labsize(medium))
```

For a smaller gap, specify negative arguments, say, `height(-1)` in the first command or `margin(0 -4 0 0)` in the second. A bit of trial and error will quickly give a satisfactory result.

Note that a sufficiently large negative argument in either `height()` or `margin()` will permit an axis title to be placed within the inner plot region, namely, inside of the axis. However, this, in turn, may cause the axis labels to disappear off the graph, so that some fiddling with the `graphregion()` option and its own `margin()` suboption may then be required (see [G-3] *region_options* and [G-4] *marginstyle*). For example,

```
. graph twoway scatter price weight,
        ytitle("Price of Cars in {c S|}US", height(-20))
        ylab(0(1000)15000, ang(hor) labsize(medium))
```

```
. graph twoway scatter price weight,
        ytitle("Price of Cars in {c S|}US", height(-20))
        ylab(0(1000)15000, ang(hor) labsize(medium))
        graphregion(margin(l+20))
```

margin() allows more flexibility in axis title positioning than does height(), but the price is a slightly more complicated syntax. For example, the $y$ axis title may be moved farther from the axis labels and closer to the top of the graph by specifying both the right-hand margin and the bottom margin of the text within the box:

```
. graph twoway scatter price weight,
        ytitle("Price of Cars in {c S|}US", margin(0 10 40 0))
        ylab(0(1000)15000, ang(hor) labsize(medium))
```

The Stata Journal (2004)
**4**, Number 3, p. 356

# Stata tip 11: The nolog option with maximum-likelihood modeling commands

Patrick Royston
MRC Clinical Trials Unit, London
patrick.royston@ctu.mrc.ac.uk

Many Stata commands fit a model by maximum likelihood, and in so doing, they include a report on the iterations of the algorithm towards (it is hoped) eventual convergence. There may be tens or even hundreds or thousands of such lines in a report, which are faithfully recorded in any log file you may have open. Suppose that you did this with stcox:

```
. use http://www.stata-press.com/data/r8/cancer, clear
. describe
. stset studytime, failure(died)
. xi: stcox i.drug age, nohr
```

You get six useless lines of output detailing the progress of the algorithm. This is a nice example, as sometimes progress is much slower or more complicated.

Those lines are of little or no statistical interest in most examples and may be omitted by adding the nolog option:

```
. xi: stcox i.drug age, nohr nolog
```

The nolog option works with many vital Stata commands, including glm, logistic, streg, and several more. My own view is that nolog should be the default in all of them. Be that as it may, you can compress your logs to good effect by specifying nolog routinely. It will remain obvious when your estimation fails to converge.

The Stata Journal (2004)
**4**, Number 3, pp. 357–358

# Stata tip 12: Tuning the plot region aspect ratio[1]

Nicholas J. Cox
University of Durham, UK
n.j.cox@durham.ac.uk

Sometimes you want a graph to have a particular shape. Graph shape is customarily quantified by the aspect ratio (height/width). One standard way of controlling the aspect ratio is setting the graph height and width by specifying the `ysize()` and `xsize()` options of `graph display`. See also [G-2] **graph display** and [G-3] *region_options*. These options control the size and, thus, the shape of the entire available graph area, including titles and other stuff beyond the plot region. At best, this is an indirect way of controlling the plot region shape, which is likely to be your main concern.

In the 23 July 2004 update, Stata 8 added an `aspect()` option to `graph` to meet this need. For example, `aspect(1)` specifies equal height and width, so that the rectangular plot region becomes a square. (A rectangle that is not a square is, strictly, an oblong.) You might want a square plot as a matter either of logic or of taste. Suppose that you are contemplating uniform random numbers falling like raindrops on the unit square within the real plane (or the plain, as the old song has it):

```
. clear
. set obs 100
. gen y = runiform()
. gen x = runiform()
. scatter y x, aspect(1) xla(0(0.1)1, grid) yla(0(0.1)1, ang(h) grid)
> yti(, orient(horiz)) plotregion(margin(zero))
```

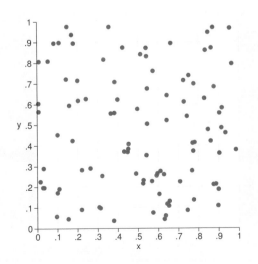

---

1. This tip was updated to reflect the function name change from `uniform()` to `runiform()`.—Ed.

In effect you have drawn a map, and maps customarily have equal vertical and horizontal distance scales or, more simply put, a single distance scale. Hence the aspect ratio is set as 1 whenever the plot region has equal extent on both axes. The same preference applies also to various special graphs on the unit square, such as ROC or Lorenz curves.

In other circumstances, the aspect ratio sought might differ from 1. Fisher (1925, 31) recommended plotting data so that lines make approximately equal angles with both axes; the same advice of banking to 45° is discussed in much more detail by Cleveland (1993). Avoiding roller-coaster plots of time series is one application. In practice, a little trial and error will be needed to balance a desire for equal angles with other considerations. For example, try variations on the following:

```
. use http://www.stata-press.com/data/r8/htourism.dta
. tsline mvdays, aspect(0.15) yla(, ang(h))
```

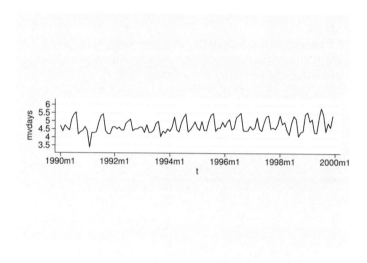

# References

Cleveland, W. S. 1993. *Visualizing Data*. Summit, NJ: Hobart.

Fisher, R. A. 1925. *Statistical Methods for Research Workers*. Edinburgh: Oliver & Boyd.

The Stata Journal (2004)
**4**, Number 4, pp. 484–485

# Stata tip 13: generate and replace use the current sort order

Roger Newson
King's College London, UK
roger.newson@kcl.ac.uk

Did you know that `generate` and `replace` use the current sort order? You might have guessed this because otherwise the `sum()` function could work as designed only with difficulty. However, this fact is not documented in the manuals, but only in the Stata web site FAQs. The consequence is that, given a particular desired `sort` order, you can be sure that values of a variable are calculated in that order and can use them to calculate subsequent values of the same variable.

A simple example is filling in missing values by copying the previous nonmissing value. The syntax for this is simply

```
. replace myvar = myvar[_n-1] if missing(myvar)
```

Here the subscript `[_n-1]`, based on the built-in variable `_n`, refers to the previous observation in the present sort order. To find more about subscripts, see [U] **13.7 Explicit subscripting** or the online help for `subscripting`.

Suppose that values of `myvar` are present for observations 1, 2, and 5 but missing in observations 3, 4, and 6. `replace` starts by replacing `myvar[3]` with the nonmissing `myvar[2]`. It then replaces `myvar[4]` with `myvar[3]`, which now contains (just in time) a copy of the nonmissing `myvar[2]`. Finally, `replace` puts a copy of `myvar[5]` into `myvar[6]`. As said, this all requires that data are in the desired sort order, commonly that of some time variable. If not, reach for the `sort` command.

There are numerous variations on this idea. Suppose that a sequence of years contains nonmissing values only for years like 1980, 1990, and 2000. This is common in data derived from spreadsheet files. A simple fix would be

```
. replace year = year[_n-1] + 1 if mi(year)
```

That way, changes cascade down the observations.

More exotic examples concern recurrence relations, as found in probability theory and elsewhere in mathematics. We typically use `generate` to define the first value (or the first few values) and `replace` to define the other values.

Consider the famous "birthday problem": what is the probability that no two out of $n$ people have the same birthday? Assuming equal probabilities of birth on each of 365 days, and so ignoring leap years and seasonal fertility variation, this probability is $\prod_{j=1}^{n} x_j$, where $x_j = (365 - j + 1)/365$. We can put these probabilities into a variable `palldiff` by typing

```
. set obs 370
. generate double palldiff = 1
. replace palldiff = palldiff[_n-1] * (365 - _n + 1) / 365 in 2/l
. label var palldiff "Pr(All birthdays are different)"
. list palldiff
```

To illustrate, the probability that all birthdays are different is below 0.5 for 23 people, below one-millionth for 97 people, and zero for over 365 people. An alternative solution (based on a suggestion by Roberto Gutierrez) is to replace the second and third lines of the above program with

```
. generate double palldiff = 0
. replace palldiff = exp(sum(ln(366 - _n) - ln(365))) in 1/365
```

which works because the product of positive numbers is the sum of their logarithms, exponentiated.

Another example is the Fibonacci sequence, defined by $y_1 = y_2 = 1$ and otherwise by $y_n = y_{n-1} + y_{n-2}$. The first 20 numbers are given by

```
. set obs 20
. generate y = 1
. replace y = y[_n-1] + y[_n-2] in 3/l
. list y
```

If you ever want to work backwards by referring to later observations, it is often easiest to reverse the order of observations and then to use tricks like these.

The Stata Journal (2004)
4, Number 4, pp. 486–487

# Stata tip 14: Using value labels in expressions

Kenneth Higbee
StataCorp
khigbee@stata.com

Did you know that there is a way in Stata to specify value labels directly in an expression, rather than through the underlying numeric value? You specify the label in double quotes (`" "`), followed by a colon (`:`), followed by the name of the value label. If we read in this dataset and see what it contains

```
. webuse census9
(1980 Census data by state)

. describe
Contains data from http://www.stata-press.com/data/r8/census9.dta
  obs:            50                          1980 Census data by state
 vars:             5                          16 Jul 2002 18:29
 size:         1,550 (99.9% of memory free)

              storage  display    value
variable name   type   format     label     variable label

state          str14   %-14s                 State
drate          float   %9.0g                 Death Rate
pop            long    %12.0gc               Population
medage         float   %9.2f                 Median age
region         byte    %-8.0g     cenreg     Census region

Sorted by:
```

we notice that variable `region` has values labeled by the `cenreg` value label. The correspondence between the underlying number and the value label is shown by

```
. label list
cenreg:
           1 NE
           2 N Cntrl
           3 South
           4 West
```

[R] **regress** uses this dataset to illustrate weighted regression. To obtain the regression of `drate` and `medage` restricted to the "South" region, you could type

```
. regress drate medage [aweight=pop] if region == 3
```

But, if you do not remember the underlying region number for "South", you could also obtain this regression by typing

```
. regress drate medage [aweight=pop] if region == "South":cenreg
(sum of wgt is   7.4734e+07)
```

| Source | SS | df | MS | | |
|---|---|---|---|---|---|
| Model | 1072.30989 | 1 | 1072.30989 | | |
| Residual | 550.163155 | 14 | 39.2973682 | | |
| Total | 1622.47305 | 15 | 108.16487 | | |

```
      Number of obs =      16
      F(  1,     14) =   27.29
      Prob > F      =  0.0001
      R-squared     =  0.6609
      Adj R-squared =  0.6367
      Root MSE      =  6.2688
```

| drate | Coef. | Std. Err. | t | P>\|t\| | [95% Conf. Interval] |
|---|---|---|---|---|---|
| medage | 3.905819 | .7477109 | 5.22 | 0.000 | 2.302139   5.509499 |
| _cons | -29.34031 | 22.33676 | -1.31 | 0.210 | -77.2479   18.56727 |

Typing the value label instead of the underlying number makes it unlikely that you
will obtain an unintended result from entering the wrong region number. An added
benefit of using the value label is that, when you later review your results, you will
quickly see that the regression is for the "South" region, and you will not need to
remember what region was assigned number 3.

See [U] **13.10 Label values** for further information about specifying value labels in
expressions.

The Stata Journal (2004)
**4**, Number 4, pp. 488–489

# Stata tip 15: Function graphs on the fly

Nicholas J. Cox
University of Durham, UK
n.j.cox@durham.ac.uk

[G-2] **graph twoway function** gives several examples of how function graphs may be drawn on the fly. The manual entry does not quite explain the full flexibility and versatility of the command. Here is a further advertisement on its behalf. To underline a key feature: you do not need to create variables following particular functions. The command handles all of that for you. We will look at two more examples.

A common simple need is to draw a circle. A trick with `twoway function` is to draw two half-circles, upper and lower, and combine them. If you are working in some scheme using color, you will usually also want to ensure that the two halves are shown in the same color. The `aspect()` option was explained in Cox (2004).

```
. twoway function sqrt(1 - x * x), ra(-1 1) ||
>        function -sqrt(1 - x * x), ra(-1 1) aspect(1)
>        legend(off) yla(, ang(h)) ytitle(, orient(horiz)) clp(solid)
```

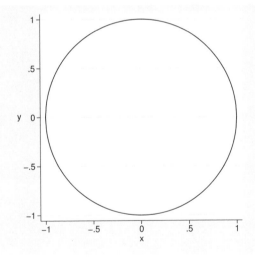

MacKay (2003, 316) asserts that, if we transform beta distributions of variables $P$ between 0 and 1 to the corresponding densities over logit $P = \ln[P/(1 - P)]$, then we find always pleasant bell-shaped densities. In contrast, densities over $P$ may have singularities at $P = 0$ and $P = 1$. This is the kind of textbook statement that should provoke some play with friendly statistical graphics software.

To explore MacKay's assertion, we need a standard result on changing variables (see, for example, Evans and Rosenthal 2004, theorems 2.6.2 and 2.6.3). Suppose that $P$ is an absolutely continuous random variable with density function $f_P$, $h$ is a function that is differentiable and monotone, and $X = h(P)$. The density function of $X$ is then

$$f_X(x) = \frac{f_P\{h^{-1}(x)\}}{|h'\{h^{-1}(x)\}|}$$

In our case, $h(P) = \text{logit } P$, so that $h^{-1}(X) = \exp(X)/\{1+\exp(X)\}$ and $h'\{h^{-1}(X)\} = \{1 + \exp(X)\}^2/\exp(X)$. In Stata terms, beta densities transformed to the logit scale are the product of `betaden(p)` or `betaden(invlogit(x))` and `exp(x)/(1+exp(x))^2`. The latter term may be recognized as a logistic density function, which always has a bell shape.

An example pair of original and transformed distributions is given by the commands below. To explore further in parameter space, you need only vary the parameters from 0.5 and 0.5 (and, if desired, to vary the range).

```
. twoway function betaden(0.5,0.5,x), ytitle(density) xtitle(p)
. twoway function betaden(0.5,0.5,invlogit(x)) * (exp(x) / (1 + exp(x))^2),
> ra(-10 10) ytitle(density) xtitle(logit p)
```

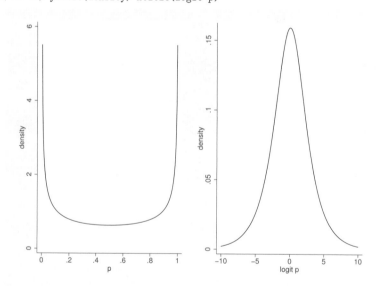

# References

Cox, N. J. 2004. Stata tip 12: Tuning the plot region aspect ratio. *Stata Journal* 4: 357–358.

Evans, M. J., and J. S. Rosenthal. 2004. *Probability and Statistics: The Science of Uncertainty*. New York: Freeman.

MacKay, D. J. C. 2003. *Information Theory, Inference, and Learning Algorithms*. Cambridge: Cambridge University Press. Also available at http://www.inference.phy.cam.ac.uk/mackay/itprnn/book.html.

The Stata Journal (2005)
**5**, Number 1, p. 134

# Stata tip 16: Using input to generate variables[1]

Ulrich Kohler
Wissenschaftszentrum Berlin für Sozialforschung
kohler@wz-berlin.de

Sometimes using `generate` is an untidy and long-winded way to generate new variables, particularly if the variable you want to create is categorical and there are many different categories. Thus rather than using

```
. gen iso3166_2 = "AT" if country == "Austria"
. replace iso3166_2 = "BE" if country == "Belgium"
. replace iso3166_2 = "TR" if country == "Turkey"
```
*and so on for say 28 countries*
```
. gen iso3166_3 = "AUT" if country == "Austria"
. replace iso3166_3 = "BEL" if country == "Belgium"
. replace iso3166_3 = "TUR" if country == "Turkey"
```
*and so on for say 28 countries*
```
. gen gername = "Österreich" if country == "Austria"
. replace gername = "Belgien" if country == "Belgium"
. replace gername = "Türkei" if country == "Turkey"
```
*and so on for say 28 countries*

you can use `input` to produce a new dataset, `save` to a temporary file, and then `merge`:

```
. preserve
. clear
. input str15 country str2 iso3166_2 str3 iso3166_3 str15 gername
Austria AT AUT Österreich
Belgium BE BEL Belgien
Turkey  TK TUR Türkei
```
*and so on*
```
. end

. sort country
. tempfile foo
. save `foo´
. restore
. sort country
. merge m:1 country using `foo´
```

Among the benefits are less typing; a cleaner log file; in huge datasets, faster data processing; and arguably fewer errors.

See [D] **input** for the finer points on `input`.

---

1. This tip was updated to reflect the new `merge` syntax.—Ed.

The Stata Journal (2005)
**5**, Number 1, pp. 135–136

# Stata tip 17: Filling in the gaps

Nicholas J. Cox
University of Durham, UK
n.j.cox@durham.ac.uk

The `fillin` command (see [D] **fillin**) does precisely one thing: it fills in the gaps in a rectangular data structure. That is very well explained in the manual entry, but people who do not yet know the command often miss it, so here is one more plug. Suppose that you have a dataset of people and the choices they make, something like this:

```
id   choice
1    1
2    3
3    1
4    2
```

Now suppose that you wish to run a nested logit model using `nlogit` (see [R] **nlogit**). This command requires all choices, those made and those not made, to be explicit. With even 4 values of `id` and 3 values of `choice`, we need 12 observations so that each combination of variables exists once in the dataset; hence, 8 more are needed in this case. The solution is just

```
. fillin id choice
```

and a new variable, `_fillin`, is added to the dataset with values 1 if the observation was "filled in" and 0 otherwise. Thus `count if _fillin` tells you how many observations were added. You will often want to `replace` or `rename _fillin` to something appropriate:

```
. rename _fillin chosen
. replace chosen = 1 - chosen
```

If you do not `rename` or `drop _fillin`, it will get in the way of a subsequent `fillin`. Usually, the decision is clear-cut: Either `_fillin` has a natural interpretation, so you want to keep it, or a relative, under a different name; or `_fillin` was just a by-product, and you can get rid of it without distress.

Another common variant is to show zero counts or amounts explicitly. With a dataset of political donations for several years, we might want an observation showing that `amount` is zero for each pair of `donor` and `year` not matched by a donation. This typically leads to neater tables and graphs and may be needed for modeling: in particular, for panel models, the zeros must be present as observations. The main idea is the same, but the aftermath is different:

```
. fillin donor year
. replace amount = 0 if _fillin
```

Naturally if we have more than one donation from various donors in various years, we might also want to `collapse` (or just possibly `contract`) the data, but that is the opposite kind of problem.

Yet another common variant is the creation of a grid for some purpose, perhaps before data entry, or before you draw a graph. You can be very lazy by typing

```
. clear
. set obs 20
. gen y = _n
. gen x = y
. fillin y x
```

which creates a $20 \times 20$ grid. This is good, but sometimes you want something different; see functions `fill()` and `seq()` in [D] **egen**.

The messiest `fillin` problems are when some of the categories you want are not present in the dataset at all. If you know a person is not one of the values of `donor`, no amount of filling in will add a set of zeros for that person. One strategy here is to add pseudo-observations so that every category occurs at least once and then to `fillin` in terms of that larger dataset. This is just a variation on the technique for creating a grid out of nothing.

As far as you can see from what is here, `fillin` just does things in place, so you need not worry about file manipulation. This is an illusion, as underneath the surface, `fillin` is firing up `cross` (see [D] **cross**), which does the work using files. Thus `cross` is more fundamental. A forthcoming tip will say more.

The Stata Journal (2005)
**5**, Number 1, pp. 137–138

# Stata tip 18: Making keys functional

Shannon Driver
StataCorp
s.driver@stata.com

Did you know that you can create custom definitions for your *F*-keys in Stata?

*F*-key definitions are created via global macros. On startup, Stata sets the *F*-key defaults to

| *F*-key | definition |
|---------|------------|
| *F1* | `help` |
| *F2* | `#review;` |
| *F3* | `describe;` |
| *F7* | `save` |
| *F8* | `use` |

You can redefine these keys if you wish.

When a definition ends with a semicolon (;), Stata will automatically execute that command as if you typed it and pressed the Enter key; otherwise, the command is immediately entered into the command line as if you had typed it. Stata then waits for you to press the Enter key. This allows you to modify the command before it is executed.

For example, to define the *F4* key to execute the `list` command, you would type

```
. global F4 "list;"
```

The "F4" here is actually a capital `F` followed by the number 4.

The best place to create these definitions is in an ASCII text file called `profile.do`. Every time Stata is launched, it looks for `profile.do` and, if it finds it, executes all of the commands it contains. For more information, type `help profile`.

Let's say that you want to create a definition for *F4* to open a window showing the contents of a particular directory. You could do this on Windows by typing

```
. global F4 `"winexec explorer C:\data;"'
```

On a Macintosh, you could type

```
. global F4 `"!open /Applications/Stata8/Stata;"'
```

You can also create *F*-key definitions to launch your favorite text editor.

```
. global F5 `"winexec notepad;"'
```

Yet another application is programming the ` and ' keys, which Stata uses to delimit local macros. Many keyboards do not have the left- or open-quote character of this pair, so an alternative is to define an *F*-key to be that key. For symmetry, you might want another *F*-key to be the right- or close-quote character. But how do you define a replacement for a key if you do not have that key in the first place? One answer lies in Stata's `char()` function:

```
. global F4 = char(96)
. global F5 = char(180)
```

You may want to make a note that *F10* is reserved internally by Windows, so you cannot program this key. Also, not all Macintosh keyboards have *F*-keys.

For more information on this topic, please see [U] **10.2 F-keys**.

The Stata Journal (2005)
5, Number 2, p. 279

# Stata tip 19: A way to leaner, faster graphs

Patrick Royston
MRC Clinical Trials Unit
p.royston@ctu.mrc.ac.uk

If you have many variables, consider doing a `preserve` of the data and `dropping` several of them before drawing a graph. This greatly speeds up production.

Take plotting fitted values from a model as an example. If there are many tied observations at each value of the predictor and therefore many replicates of the fitted values, the size of the graph file can be large, also making the plotting time large. A construction like this can save resources:

```
. preserve
. bysort x: drop if _n > 1
. line f1 f2 f3 x, sort clp(1 - _) saving(graph, replace)
. restore
```

Here is another real example: with 15,156 variables and 50 observations, I wanted a `dotplot` of variable v15155 by v15156. The time taken with all data present was 10.66 seconds, but with `preserve` and all irrelevant variables `dropped`, it was 0.69 seconds.

The Stata Journal (2005)
**5**, Number 2, pp. 280–281

# Stata tip 20: Generating histogram bin variables[1]

David A. Harrison
ICNARC, London, UK
david.harrison@icnarc.org

Did you know about `twoway__histogram_gen`? (Note the two underscores in the first gap and only one in the second.) This command is used by `histogram` to generate the variables that are plotted. It is undocumented in the manuals but explained in the online help. The command can be used directly to save these variables, enabling more complex manipulation of histograms and production of other graphs or tables.

Consider the S&P 500 historical data that are used as an example for [R] **histogram**:

```
. use http://www.stata-press.com/data/r9/sp500
(S&P 500)
. histogram volume, percent start(4000) width(1000)
(bin=20, start=4000, width=1000)
    (output omitted)
```

To display only the central part of this histogram from 8,000 to 16,000, we could use `if`, but this will change the height of the bars, as data outside the range 8,000 to 16,000 will be ignored completely. To restrict the range without altering the bars, we use `twoway__histogram_gen` to save the histogram and only plot the section of interest:

```
. twoway__histogram_gen volume, percent start(4000) width(1000) gen(h x)
. twoway bar h x if inrange(x,8000,16000), barwidth(1000) bstyle(histogram)
```

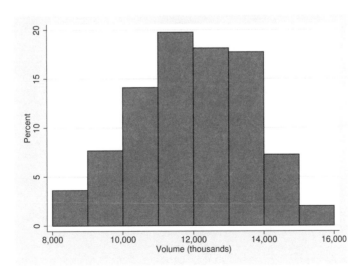

---

1. This tip was updated to reflect the function name change from `norm()` to `normal()`.—Ed.

The `start()` and `width()` options above specified cutpoints that included 8,000 and 16,000. We could, alternatively, use the default cutpoints:

```
. twoway__histogram_gen volume if inrange(volume,8000,16000), display
(bin=14, start=8117, width=525.08571)
. local m = r(start)
. local w = r(width)
. summarize volume, meanonly
. local s = `m' - `w' * ceil((`m' - r(min))/`w')
. twoway__histogram_gen volume, percent start(`s') width(`w') gen(h x, replace)
. twoway bar h x if inrange(x,8000,16000), barwidth(`w') bstyle(histogram)
  (output omitted)
```

Other uses of `twoway__histogram_gen` include the following:

- Overlaying or mirroring two histograms

```
. use http://www.stata-press.com/data/r9/bplong, clear
(fictional blood-pressure data)
. twoway__histogram_gen bp if sex == 0, frac start(125) w(5) gen(h1 x1)
. twoway__histogram_gen bp if sex == 1, frac start(125) w(5) gen(h2 x2)
. twoway (bar h1 x1, barw(5) bc(gs11))
> (bar h2 x2, barw(5) blc(black) bfc(none)),
> legend(order(1 "Male" 2 "Female"))
  (output omitted)
. qui replace h2 = -h2
. twoway (bar h1 x1, barw(5)) (bar h2 x2, barw(5)),
> yla(-.2 ".2" -.1 ".1" 0 .1 .2) legend(order(1 "Male" 2 "Female"))
  (output omitted)
```

- Changing the scale, for example, to plot density on a square-root scale

```
. twoway__histogram_gen bp, start(125) width(5) gen(h x)
. qui gen hsqrt = sqrt(h)
. twoway bar hsqrt x, barw(5) bstyle(histogram) ytitle(Density)
> ylabel(0 .05 ".0025" .1 ".01" .15 ".0225" .2 ".04")
  (output omitted)
```

- Plotting the differences between observed and expected frequencies

```
. twoway__histogram_gen bp, freq start(125) w(5) gen(h x, replace)
. qui summarize bp
. qui gen diff = h - r(N) * (normal((x + 2.5 - r(mean))/r(sd)) -
> normal((x - 2.5 - r(mean))/r(sd)))
. twoway bar diff x, barw(5) yti("Observed - expected frequency")
  (output omitted)
```

There are also two similar commands: `twoway__function_gen` to generate functions and `twoway__kdensity_gen` to generate kernel densities.

The Stata Journal (2005)
**5**, Number 2, pp. 282–284

# Stata tip 21: The arrows of outrageous fortune

Nicholas J. Cox
Durham University
n.j.cox@durham.ac.uk

Stata 9 introduces a clutch of new plottypes for `graph twoway` for paired-coordinate data. These are defined by four variables, two specifying starting coordinates and the other two specifying ending coordinates. Here we look at some of the possibilities opened up by [G-2] **graph twoway pcarrow** for graphing changes over time. Arrows are readily understood by novices as well as experts as indicating, in this case, the flow from the past towards the present.

Let us begin with one of the classic time-series datasets. The number of lynx trapped in an area of Canada provides an excellent example of cyclic boom-and-bust population dynamics. Trappings are optimistically assumed to be proportional to the unknown population size.

The dataset has already been `tsset` (see [TS] **tsset**).

```
. use http://www.stata-press.com/data/r9/lynx2.dta, clear
(TIMESLAB: Canadian lynx)

. tsline lynx
  (output omitted)
```

Ecologists and other statistically minded people find it natural to think about populations on a logarithmic scale: population growth is after all inherently multiplicative. Logarithms to base 10 are convenient for graphing.

```
. gen loglynx = log10(lynx)

. twoway pcarrow loglynx L.loglynx F.loglynx loglynx,
> xla(2 "100" `=log10(200)´ "200" `=log10(500)´ "500" 3 "1000" `=log10(2000)´
> "2000" `=log10(5000)´ "5000")
> yla(2 "100" `=log10(200)´ "200" `=log10(500)´ "500" 3 "1000" `=log10(2000)´
> "2000" `=log10(5000)´ "5000")
> ytitle(this year) xtitle(previous year) subtitle(Number of lynx trapped)
```

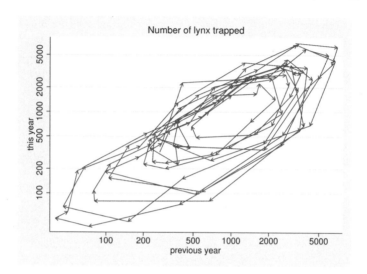

Thinking as it were autoregressively, we can plot this year's population versus the previous year's and join data points with arrows end to end. Each data point other than the first and last is the end of one arrow pointing from L.loglynx to loglynx and the beginning of another pointing from loglynx to F.loglynx. This dual role and the ability to use time-series operators such as L. and F. on the fly in graphics commands yield the command syntax just given. Plotted as a trajectory in this space, population cycles are revealed clearly as asymmetric. Depending on your background, you may see this as an example of hysteresis, or whatever else it is called in your tribal jargon.

Another basic comparison compares values for some outcome of interest at two dates. For this next example, we use life expectancy data for 1970 and 2003 from the UNICEF report, *The State of the World's Children 2005*, taken from the web site *http://www.unicef.org* accessed on May 12, 2005. A manageable graph focuses on those countries for which life expectancy was under 50 years in 2003. A count on the dataset thus entered shows that there are 33 such countries.

We borrow some ideas from displays possible with **graph dot** (see [G-2] **graph dot**). Arrows connecting pairs of variables are not supported by **graph dot**. However, as is common with Stata's graphics, whatever is difficult with **graph dot**, **graph bar**, or **graph hbar** is often straightforward with **graph twoway**, modulo some persistence.

A natural sort order for the graph is that of life expectancy in 2003. A nuance to make the graph tidier is to break ties according to life expectancy in 1970. Life expectancy is customarily, and sensibly, reported in integer years, so ties are common. One axis for the graph is then just the observation number given the sort order, except that we will want to name the countries concerned on the graph. For names that might be fairly long, we prefer horizontal alignment and thus a vertical axis. The names are best assigned to value labels. Looping over observations is one way to define those. The online help on `forvalues` and `macros` explains any trickery with the loop that may be unfamiliar to you; also see [P] **forvalues** and [P] **macro**.

```
. gsort lifeexp03 - lifeexp70
. gen order = _n
. forval i = 1/33 {
  2.          label def order `i´ "`=country[`i´]´", modify
  3. }
. label val order order
```

The main part of the graph is then obtained by a call to `twoway pcarrow`. The arrowhead denotes life expectancy in 2003. Optionally, although not essentially, we overlay a scatter plot showing the 1970 values.

```
. twoway pcarrow order lifeexp70 order lifeexp03 if lifeexp03 < 50
> || scatter order lifeexp70 if lifeexp03 < 50, ms(oh)
> yla(1/33, ang(h) notick valuelabel labsize(*0.75)) yti("") legend(off)
> barbsize(2) xtitle("Life expectancy in years, 1970 and 2003") aspect(1)
```

Apart from Afghanistan, all the countries shown are in Africa. Some show considerable improvements over this period, but in about as many, life expectancy has fallen dramatically. Readers can add their own somber commentary in terms of war, political instability, famine, and disease, particularly AIDS.

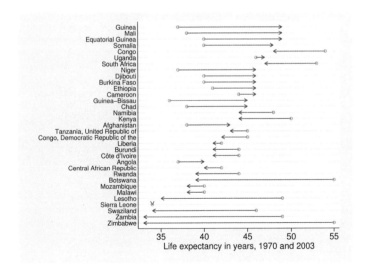

The Stata Journal (2005)
**5**, Number 3, pp. 465–466

36

# Stata tip 22: Variable name abbreviation

Philip Ryan
University of Adelaide
philip.ryan@adelaide.edu.au

Stata allows users to abbreviate any variable name to the shortest string of characters that uniquely identifies it, given the data currently loaded in memory (see [U] **11.2.3 Variable-name abbreviation**). Stata also offers three wildcard characters, *, ~, and ? (see [U] **11.4.1 Lists of existing variables**), so users have substantial flexibility in how variables may be referenced.

Stata also allows users to control the values of many of its system parameters using the `set` command (see [R] **set**). One of these parameters is `varabbrev`, which may be toggled on, allowing variable names to be abbreviated, or off, requiring the user to spell out entire variable names.

The default is to allow abbreviations. But this convenience feature can bite. Suppose that in a program we wish to confirm the existence of a variable and that variable does not in fact exist:

```
. clear
. set varabbrev on
. set obs 10
(obs was 0, now 10)
. generate byte myvar7 = 1
. confirm variable myvar
```

There is no error message here because `myvar` is an allowed abbreviation for `myvar7`. A bigger deal is that as `myvar7` exists and not `myvar`, typing `drop myvar` would drop `myvar7`, which may or may not have been our intention.

But what if we had wanted to confirm explicitly the existence of variable `myvar`? There are two ways to do this:

1. Specify the `confirm` command with the `exact` option (see [P] **confirm**):

   ```
   . confirm variable myvar, exact
   variable myvar not found
   r(111);
   ```

2. Toggle variable abbreviation off:

   ```
   . set varabbrev off
   . confirm variable myvar
   variable myvar not found
   r(111);
   ```

Note that the status of `varabbrev` does not affect the display of variable names. For example,

```
. sysuse auto, clear
(1978 Automobile Data)
. set varabbrev off
. rename weight this_is_a_very_long_varname
. regress price turn length this_is_a_very_long_varname
```
(*output omitted*)

| price | Coef. | Std. Err. | t | P>\|t\| | [95% Conf. Interval] | |
|---|---|---|---|---|---|---|
| turn | -318.2055 | 127.1241 | -2.50 | 0.015 | -571.7465 | -64.66452 |
| length | -66.17856 | 39.87361 | -1.66 | 0.101 | -145.704 | 13.34684 |
| this_is_a_~e | 5.382135 | 1.116756 | 4.82 | 0.000 | 3.154834 | 7.609435 |
| _cons | 14967.64 | 4541.836 | 3.30 | 0.002 | 5909.228 | 24026.04 |

In this display, Stata has abbreviated the long variable name, despite the current value of `varabbrev`.

Note that the `list` command has its own option to allow the user partial control of the display; see [D] **list**. As we `set varabbrev off`, we must specify only unabbreviated variable names in a `list` command, but we can override Stata's default abbreviation in the display using the `abbreviate()` option:

```
. list make turn this_is_a_very_long_varname in 1/4, abb(21)
```

| | make | turn | this_is_a_very_long~e |
|---|---|---|---|
| 1. | AMC Concord | 40 | 2,930 |
| 2. | AMC Pacer | 40 | 3,350 |
| 3. | AMC Spirit | 35 | 2,640 |
| 4. | Buick Century | 40 | 3,250 |

The default value for `abbreviate()` is 8, so that otherwise the variable name would have been displayed as `this_i~e`.

The Stata Journal (2005)
5, Number 3, pp. 467–468

# Stata tip 23: Regaining control over axis ranges

Nicholas J. G. Winter
Cornell University
nw53@cornell.edu

Beginning with version 8, Stata will often widen the range of a graph axis beyond the range of the data. Convincing Stata to narrow the range can be difficult unless you understand the cause of the problem.

Using the trusty `auto` dataset, consider the graph produced by this command:

```
. sysuse auto, clear
(1978 Automobile Data)

. twoway scatter mpg price
```

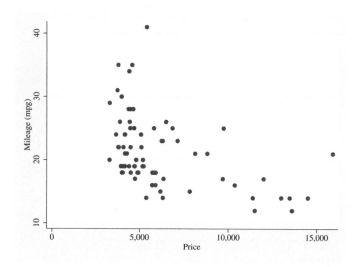

Although price ranges from \$3,291 to \$15,906 in the data, the lower end of the $x$-axis in this graph extends to zero, leaving blank space on the left-hand side. If we do not like this space, the solution would seem to lie with the `range()` suboption of the `xscale()` option. Thus we might expect Stata to range the $x$-axis using the minimum and maximum of the data, given the following command:

```
. twoway scatter mpg price, xscale(range(3291 15906))
```

However, this produces the same graph: the axis still includes zero. It seems that Stata is ignoring `range()`, although it does not do that when the range is *increased*, rather than decreased. Consider, for example, this command, which expands the $x$-axis to run from 0 through 30,000:

```
. twoway scatter mpg price, xscale(range(0 30000))
```

The issue is that the range displayed for an axis depends on the interaction between two sets of options (or their defaults): those that control the axis range explicitly, and those that *label* the axis. The range can be expanded either by explicitly specifying a longer axis (e.g., with `xscale(range(a b))`) or by labeling values outside the range of the data.

To determine the range of an axis, Stata begins with the minimum and maximum of the data. Then it will widen (but never narrow) the axis range as instructed by `range()`. Finally, it will widen the axis if necessary to accommodate any axis labels.

By default, `twoway` labels the axes with "about" five ticks, the equivalent of specifying `xlabel(#5)`. In this case, Stata chooses four labels, one of which is zero, and then expands the $x$-axis accordingly. In other words, if we specify `xscale()`—but do not specify `xlabel()`—we are in effect saying to Stata "and please use the default `xlabel()` for this graph". This default may widen the axis range.

Therefore, to get a narrower $x$-axis, we must specify a narrower set of axis labels. For example, to label just the minimum and maximum, we could specify

```
. scatter mpg price, xlabel(minmax)
```

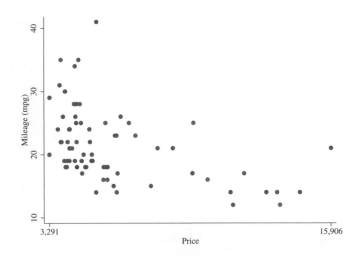

Of course, we could specify any other set of points to label, for example,

```
. scatter mpg price, xlabel(5000[1000]15000)
```

This issue did not appear until Stata version 8. Prior versions defaulted to labeling the minimum and the maximum of the data only. This would be equivalent to including the options `xlabel(minmax)` and `ylabel(minmax)` in Stata 8 or later.

The Stata Journal (2005)
**5**, Number 3, p. 469

# Stata tip 24: Axis labels on two or more levels

Nicholas J. Cox
Durham University
n.j.cox@durham.ac.uk

Text shown as graph axis labels is by default shown on one level. For example, a label `Foreign cars` would be shown just like that. Sometimes you want the text of a label to be shown on two or even more levels, as one way of reducing crowding or even overprinting of text; thus you might want `Foreign` written above `cars`. Other ways of fighting crowding include varying the size or angle at which text is printed (see [G-3] *axis_label_options* for details), or in some cases reconsidering which variable should go on which axis.

To specify multiple levels, the text to go on each level should appear within double quotes " ", and the whole text label should appear within compound double quotes ' " " '. For more explanation of the latter, see [U] **18.3.5 Double quotes**. That way, Stata's parser has a clear idea of parts and wholes.

Here are some examples:

```
. sysuse auto
. dotplot mpg, over(foreign)
> xlabel(0 `" "Domestic" "cars" "' 1 `" "Foreign" "cars" "') xtitle("")
. graph box mpg,
> over(foreign, relabel(1 `" "Domestic" "cars" "'  2 `" "Foreign" "cars" "'))
. graph hbar (mean) mpg,
> over(foreign, relabel(1 `" "Domestic" "cars" "'  2 `" "Foreign" "cars" "'))
```

Note the subtle difference between these examples. `dotplot` is really a wrapper for `twoway` and, as is characteristic of `twoway` graphs, it takes its variables literally so that the values of `foreign` are indeed treated as 0 and 1. On the other hand, graphs with so-called categorical axes (`graph bar`, `graph hbar`, `graph box`, `graph hbox`, and `graph dot`) consider the categories shown to be 1, 2, and so forth, regardless of the precise numeric or string values of the variables concerned. The numbers increase from left to right or from top to bottom, as the case may be. Thus matrix users will feel at home with this convention.

The Stata Journal (2005)
**5**, Number 4, pp. 602–603

# Stata tip 25: Sequence index plots

Ulrich Kohler and Christian Brzinsky-Fay
Wissenschaftszentrum Berlin
kohler@wz-berlin.de, brzinsky-fay@wz-berlin.de

Sequence index plots of longitudinal or panel data use stacked bars or line segments to show how individuals move between a set of conditions or states over time. Changes of state are shown by changes of color. The term *sequence index plot* was proposed by Brüderl and Scherer (2004). See Scherer (2001) for an application.

It is possible to draw sequence index plots with Stata by using the `twoway` plottype `rbar`. Starting from data in survival-time form (see `help st`), you simply overlay separate range-bar plots for each state.

For example, suppose that you have data on times for entering and leaving various states of employment:

```
. list in 1/10
```

|  | id | type | begin | end |
|---|---|---|---|---|
| 1. | 1 | employed | 1 | 13 |
| 2. | 1 | apprenticeship | 13 | 20 |
| 3. | 1 | unemployed | 20 | 23 |
| 4. | 1 | employed | 23 | 25 |
| 5. | 1 | unemployed | 25 | 26 |
| 6. | 1 | employed | 26 | 43 |
| 7. | 1 | unemployed | 43 | 50 |
| 8. | 1 | employed | 50 | 60 |
| 9. | 2 | employed | 1 | 13 |
| 10. | 2 | apprenticeship | 13 | 21 |

First, **separate** the start and end dates for the different states:

```
. separate begin, by(type)
. separate end, by(type)
```

Then plot overlaid range bars for each state:

```
. graph twoway
>    (rbar begin1 end1 id, horizontal)
>    (rbar begin2 end2 id, horizontal)
>    (rbar begin3 end3 id, horizontal)
>    (rbar begin4 end4 id, horizontal)
>    (rbar begin5 end5 id, horizontal)
>    , legend(order(1 "education" 2 "apprenticeship"
>            3 "employment" 4 "unemployment"  5 "inactivity")
>            cols(1) pos(2) symxsize(5))
>      xtitle("months") yla(, angle(h)) yscale(reverse)
```

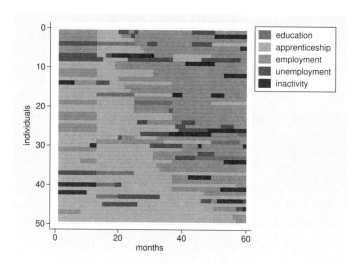

It is common to put personal identifiers on the $y$-axis, using the option `horizontal`, and put time on the $x$-axis.

In practice, with many individuals in a large panel, the bars become thinner lines. In such cases, you could use the plottype `rspike` instead. Note also that you can make room for more individuals by tuning the aspect ratio of the graph (see Cox 2004). There is no upper limit to how many individuals are shown, although as the number increases, the resulting graph may become too difficult to interpret. The readability, however, largely depends on how far similar individuals are grouped together. The sort order should therefore be some criterion of similarity between sequences.

To fine tune the graph, use any option allowed with `graph twoway`; type `help twoway_options`. Our example provides some simple illustrations. `legend()` changes the contents and placement of the legend. `xtitle()` defines the title along the $x$-axis. `ylabel()` is used to display the $y$-axis labels horizontally, instead of vertically. `yscale(reverse)` reverses the scale of the $y$-axis so that the first individual is plotted at the very top of the graph.

# References

Brüderl, J., and S. Scherer. 2004. Methoden zur Analyse von Sequenzdaten. *Kölner Zeitschrift für Soziologie und Sozialpsychologie* Sonderheft 44: Methoden der Sozialforschung: 330–347.

Cox, N. J. 2004. Stata tip 12: Tuning the plot region aspect ratio. *Stata Journal* 4: 357–358.

Scherer, S. 2001. Early career patterns: A comparison of Great Britain and West Germany. *European Sociological Review* 17: 119–144.

The Stata Journal (2005)
**5**, Number 4, p. 604

# Stata tip 26: Maximizing compatibility between Macintosh and Windows

Michael S. Hanson
Wesleyan University
mshanson@wesleyan.edu

A questioner on Statalist asked whether there are problems using Stata in a joint project on different operating systems. My short answer is "No". Underlying this is StataCorp's work to ensure that its official filetypes (`.dta`, `.gph`, `.ado`, `.do`, `.dct`, etc.) are completely compatible across all the operating systems it supports and that Stata can always read those files even from older versions.

My slightly longer answer is "Not with a few basic precautions", but much depends on how the collaboration occurs. Here is some advice for easier joint use of Stata across platforms (and indeed on the same platform).

**Standardize Statas.** Ideally, everyone should use the same version of Stata, with the same additional `.ado` files installed.

**Avoid absolute file paths.** (This strikes me as generally a good idea anyway, as it leads to greater portability.) If the project is sufficiently complex, create an identical subdirectory structure below some common project root directory on every file system and only use path references relative to this root.

**Use forward slashes in file paths.** Note that Stata understands the forward slash (`/`) as separating directory levels on all platforms, even Windows. So use that instead of the backward slash (`\`) for paths in all `.do` files.

**Watch end-of-line delimiters.** Text files have different line endings on Macintosh, Windows, and Unix systems. So long as users on different platforms are using sufficiently versatile text editors, it should be straightforward to read both input files (e.g., `.do` files) and output files (e.g., `.log` files) regardless of the line endings used—and, if necessary, to convert to the desired one. Note that how the files will be shared—common server, (S)FTP, email—may have implications for this issue.

**Use Encapsulated PostScript.** Save graphics in EPS format. (See the help or manual entry for `graph export`.) While the Macintosh can natively generate graphics in PDF format, the PC cannot (without jumping through some hoops and purchasing Acrobat, that is). WMF, EMF, and PICT do not translate well across platforms, and while you could use PNG, nonvector formats do not scale well. (As they are at a fixed size, when enlarged, they still contain only the lower-density information of the original, and when they are reduced, information must be lost to reduce their size.)

**Use other open-standards file formats.** More generally, collaboration will be easier if everyone uses open-standards file formats (e.g., plain text, PNG, TeX, or LaTeX, etc.)—instead of those tied to proprietary software.

The Stata Journal (2005)
5, Number 4, pp. 605–607

# Stata tip 27: Classifying data points on scatter plots

Nicholas J. Cox
Durham University
n.j.cox@durham.ac.uk

When you have scatter plots of counted or measured variables, you may often wish to classify data points according to the values of a further categorical variable. There are several ways to do this. Here we focus on the use of `separate`, gray-scale gradation, and text characters as class symbols. If different categories really do plot as distinct clusters, it should not matter too much how you show them, but knowing some Stata tricks should also help.

One starting point is that differing markers may be used on the plot whenever there are several variables plotted on the $y$-axis. With the `auto.dta` dataset, you can imagine

```
. sysuse auto
. gen mpg0 = mpg if foreign == 0
. gen mpg1 = mpg if foreign == 1
. scatter mpg? weight
```

Note the use of the wildcard `mpg?`, which picks up any variable names that have `mpg` followed by just one other character. Once the two variables `mpg0` and `mpg1` have been generated, different markers are automatic. This process still raises two questions. To get an acceptable graph, we need self-explanatory variable labels or at least self-explanatory text in the graph legend. Moreover, two categories are easy enough, but do we have to do this for each of say 5, 7, or 9 categories?

In fact, it would have been better to type

```
. separate mpg, by(foreign) veryshortlabel
. scatter mpg? weight
```

The command `separate` (see [D] **separate**) generates all the variables we need in one command and has a stab at labeling them intelligibly. In this case, we use the (undocumented) `veryshortlabel` option, which was implemented with graphics especially in mind. You may prefer the results of the documented `shortlabel` option. Note that the `by()` option can take true-or-false conditions, such as `price < 6000`, as well as categorical variables.

If your categorical variable consists of qualitatively different categories, you are likely to want to use qualitatively different symbols. Alternatively, if that variable is ordered or graded, the coding you use should also be ordered. One possibility is to use symbols colored in a sequence of gray scales.

Some data on landforms illustrate the point: Ian S. Evans kindly supplied measurements of 260 cirques in Wales, armchair-shaped hollows formerly occupied by small glaciers. Length tends to increase with width, approximately as a power function, but qualitative aspects of form, particularly how closely they approach a classic, well-developed shape, are also coded in a grade variable.

```
. separate length, by(grade) veryshortlabel
. scatter length? width, xsc(log) ysc(log) ms(O ..)
> mcolor(gs1 gs4 gs7 gs10 gs13) mlcolor(black ..) msize(*1.5 ..)
> yti("`: variable label length´") yla(200 500 1000 2000, ang(h))
> xla(200 500 1000 2000) legend(pos(11) ring(0) col(1))
```

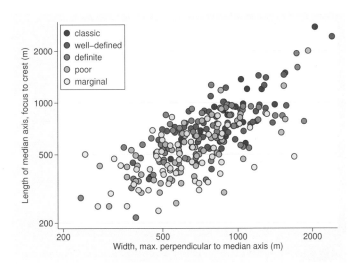

Figure 1 shows length versus width, subdivided by grade. Some practical details deserve emphasis. Gray scales near 16 (white) may be difficult to spot against a light background, including any printed page. Therefore, a dark outline color is recommended. Bigger symbols than the default are needed to do the coloring justice, but as a consequence, this approach is less likely to be useful with thousands of data points. A `by()` option showing different categories separately might work better. With the coding here, it so happens that the darkest category is plotted first and is thus liable to be overplotted by lighter categories wherever data points are dense. Some experimentation with the opposite order of plotting might be a good idea to see which works better.

An alternative that sometimes works nicely is to use ordinary text characters as different markers. One clean style is to suppress the marker symbols completely, using instead the contents of a `str1` variable as marker labels. Whittaker (1975, 224) gave data on net primary productivity and biomass density for various ecosystem types. Figure 2 shows the subdivision.

```
. scatter npp bd, xsc(log) ysc(log) ms(i) mlabpos(0) mlabsize(*1.4)
> mla(c) yla(3000 1000 300 100 30 10 3, nogrid ang(h))
> xla(0.01 "0.01" 0.1 "0.1" 1 10 100)
> legend(on ring(0) pos(5) order( - "m marine" - "w wet" - "c cultivated" -
> "g grassland" - "f forest" - "b bare"))
```

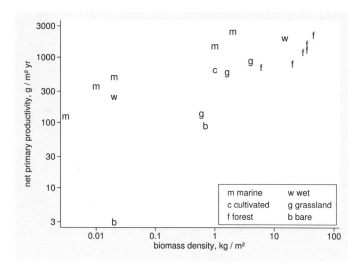

With three or four orders of magnitude variation in each variable, log scales are advisable. On those scales, there is a broad correlation whereby more biomass means higher productivity, but also considerable variation, much of which can be rationalized in terms of very different cover types. For the same biomass density, marine and other wet ecosystems have higher productivity than land ecosystems.

On the Stata side, remember `mlabpos(0)` and note that the `legend` must be set `on` explicitly. For different purposes, or for different tastes, what is here given as the legend might go better as text in a caption in a printed report. Behind the practice here lies general advice that lowercase letters, such as `abc`, work better than uppercase, such as `ABC`, as they are easier to distinguish from each other, and they are less likely to impart an synaesthetic sense in readers that the graph designer is shouting at them.

# Reference

Whittaker, R. H. 1975. *Communities and Ecosystems*. New York: Macmillan.

The Stata Journal (2006)
**6**, Number 1, pp. 144–146

# Stata tip 28: Precise control of dataset sort order

L. Philip Schumm
Department of Health Studies
University of Chicago
Chicago, IL
pschumm@uchicago.edu

The observations in a Stata dataset are ordered, so that they may be referred to by their position (e.g., in 42/48) and that individual values of a variable may be referred to with subscripts (e.g., mpg[42]). This order can be changed by using the sort command (see [D] **sort**). Developing a full appreciation of what is possible using sort together with the by: prefix, the underscore built-ins _n and _N, and subscripting is a major step toward Stata enlightenment (e.g., see Cox [2002]).

One source of surprise for many users arises when sorting by one or more variables which, when taken together, do not uniquely determine the order of observations. In this case, the resulting order within any group of observations having the same value(s) of those variables is effectively random because sort uses an *unstable* sort algorithm. Users who desire a *stable* sort—in which the previous ordering of observations within tied values of the sort variables is maintained—should specify the **stable** option. However, this option will slow sort down and, more importantly, can hide problems in your code.

You are likely to discover this issue when coding an operation dependent on the order of the data that gives different results from one run to another. Consider the following dataset consisting of mothers and their children:

```
. list, sepby(family)
```

|     | family | name | child |
|-----|--------|------|-------|
| 1.  | 2 | Harriet | 0 |
| 2.  | 2 | Lewis | 1 |
| 3.  | 1 | Sylvia | 0 |
| 4.  | 1 | Jenny | 1 |
| 5.  | 3 | Kim | 0 |
| 6.  | 3 | Peter | 1 |
| 7.  | 3 | Kim | 1 |

Individuals are grouped by family, the mother always appearing first. Suppose that we want to construct a unique within-family identifier, such that all mothers have the same value. This is a straightforward application of by:, but first the data must be sorted by family:

```
. sort family
. by family: generate individual = _n
```

```
. table child individual
```

| child | individual 1 | 2 | 3 |
|---|---|---|---|
| 0 | 2 | 1 | |
| 1 | 1 | 2 | 1 |

Unfortunately, the result is not as desired: one mother was assigned the value 2. In fact, following the call to `sort`, the order of observations within families—and hence the assignment of identifiers—was random. If we had instead sorted by family *and* child, each mother would have appeared first and would have been assigned a value of 1 (assuming that each family has exactly one mother—a key assumption that should always be checked). Yet even this solution would still be deficient: if a family has multiple children, their identifiers would be random and irreproducible. Only if we sort by family, child, *and* name would we have an adequate solution.

If we had used instead

```
. sort family, stable
```

we would also have obtained the desired result. So why does `sort` by default perform an unstable sort? Apart from better performance, the answer (emphasized by William Gould on Statalist) is that using the `stable` option not only fails to address the problem; it also reduces the chance of discovering it. Our error was to perform a calculation dependent on the sort order of the data without establishing that order beforehand. Using `stable` would have temporarily masked the error. However, had the sort order of the input dataset changed, we would have been in trouble.

How can you avoid such problems? First, train yourself to recognize when a calculation depends on the sort order of the data. Most instances in which you are using _n and _N or subscripting (either alone or with `by`) are easy to recognize. However, instances in which you are using a function that depends on the order of the data (e.g., `sum()` or `group()`) can be more subtle (Gould 2000).

Second, ensure that the order of the data is fully specified. This check became much easier in Stata 8 with the introduction of the `isid` command ([D] **isid**), which checks whether one or more variables uniquely identify the observations and returns an error if they do not. The command also has a `sort` option, which sorts the dataset in order of the specified variable(s). This option lets us replace our original `sort` command with

```
. isid family child name, sort
```

which, since it runs without error, confirms that we have specified the order fully. Had we used only `family`, or `family` and `child`, `isid` would have returned an error, immediately alerting us to the problem.

# References

Cox, N. J. 2002. Speaking Stata: How to move step by: step. *Stata Journal* 2: 86–102.

Gould, W. 2000. FAQ: Sorting on categorical variables.
http://www.stata.com/support/faqs/lang/sort.html.

The Stata Journal (2006)
**6**, Number 1, pp. 146–148

# Stata tip 29: For all times and all places

Charles H. Franklin
Department of Political Science
University of Wisconsin–Madison
Madison, WI
chfrankl@wisc.edu

According to the *Data Management Reference Manual*, the `cross` command is "rarely used"; see [D] **cross**. This comment understates the command's usefulness. For example, the `fillin` command uses `cross` (Cox 2005). Here is one further circumstance in which it proves extremely useful, allowing a simple solution to an otherwise awkward problem.

In pooled time-series cross-sectional data, we require that some number of units (geographic locations, patients, television markets) be observed over some period (daily from March to November, say). We thus need a data structure in which each unit is represented at each time point. If the data come in this complete form, then no problem arises. But when aggregating from lower-level observations, some dates, and possibly some units, are often missing. This missingness could be because no measurement was taken or because an event that is being counted simply did not occur on that date and so no record or observation was generated. In the aggregated Stata data file, no observation will appear for these dates or units. Inserting observations for the missing dates or units is awkward, but the `cross` command, followed by `merge`, makes the solution simple.

To illustrate with a real example: in the Wisconsin Advertising Project, we have coded 1.06 million political advertisements broadcast during the 2004 U.S. presidential campaign, using data provided by Nielsen Monitor-Plus. These ads are distributed across 210 media markets. Each time an ad is broadcast, it generates an observation in our dataset. The data are then aggregated to the media market to produce a daily count of the total advertising in each market. Such aggregation is simple in Stata. Variables `repad` and `demad` are coded 1 if the ad supported the Republican or Democratic candidate, respectively, and 0 otherwise. The sum is thus simply the count of the number of ads supporting each candidate.

```
clear
use allads
sort market date
collapse (sum) repad demad, by(market date) fast
save marketcounts, replace
```

This do-file produced no observation if no ads ran in a market on a particular date, which is common in these data. We want a dataset that includes every date for each of the 210 markets, with a value of 0 if no ad ran in a market on a date.

We can use `cross` to create a dataset that has one observation for each market for each of the 245 days included in our study. The file `dmacodelist.dta` contains one observation for each of the 210 markets: `dma` stands for "designated market area", Nielsen's term for television markets. First, we create a Stata dataset with 245 observations, one for each day of our study (March 3–November 2). Then we convert this information to a Stata date.

```
clear
set obs 245
gen date = _n + mdy(03,02,2004)
format date %d
```

Now use `cross` to generate the dataset with all dates for all markets:

```
cross using dmacodelist
sort market date
save alldates, replace
```

The file `alldates.dta` contains one observation for each market and for each date. The last step is to merge the aggregated `marketcount.dta` dataset with `alldates.dta` and replace missing values with zeros.

```
clear
use marketcounts
sort market date
merge market date using alldates
assert _merge != 1
replace demad = 0 if demad == .
replace repad = 0 if repad == .
```

The merge should produce no values of `_merge` that are 1, meaning observations found only in `marketcounts`, so the `assert` command checks this: the do-file will stop if the assertion is false (see Gould 2003 on `assert`). The `repads` and `demads` will be missing in the merged data only if no ad was broadcast, so replacing missing values for these variables with zeros will result in the desired dataset.

Thus the `cross` command offers an efficient solution to this type of problem. Those who often aggregate low-level data to create time-series cross-sectional structures will find this command handy.

# References

Cox, N. J. 2005. Stata tip 17: Filling in the gaps. *Stata Journal* 5: 135–136.

Gould, W. 2003. Stata tip 3: How to be assertive. *Stata Journal* 3: 448.

The Stata Journal (2006)
**6**, Number 1, pp. 149–150

# Stata tip 30: May the source be with you

Nicholas J. Cox
Department of Geography
Durham University
Durham City, UK
n.j.cox@durham.ac.uk

Stata 9 introduced a command, `viewsource`, that does two things: it finds a text file along your ado-path and then opens the Viewer on that text file. See [P] **viewsource** and [U] **17.5 Where does Stata look for ado-files?** for the basic explanations. Naturally, if the text file does not exist as named, say, because you mistyped the name or because it really does not exist, you will get an error message.

Although intended primarily for Stata programmers, `viewsource` can be useful for examining (but not for editing) any text file you are working with. That can include program files, help files, text data files, do-files, log files, and other text documents. Binary or proprietary-format files are not banned, but the command is unlikely to be useful with them. The Viewer is, in particular, not a substitute for any word processor.

Here are some examples. A good way to learn about any Stata command defined by an ado-file is to look at the source code. You might be puzzled by some output, suspect a bug, or simply be curious. Even if you are not (yet) a Stata programmer, you can learn a lot by looking at the code. After all, it is just more Stata. Many, but not all, commands, generically *cmdname*, are defined by *cmdname*.`ado`—an ado-file with the same name as the command. The exceptions are part of the executable and not visible to you by using `viewsource` or indeed any other command. You might as well start reflexively:

```
. viewsource viewsource.ado
```

from which you will see that `viewsource`'s main actions are to (try to) find the file you specified using `findfile` and then to open the file using `view`. You will see other details too, and puzzling out what they do is a good exercise in program appreciation.

A second example is opening a help file. This action may seem redundant given the existence of the `help` command, but there is a noteworthy exception. `viewsource` fires up `view` with its `asis` option, so that interpreting the SMCL commands in any help file is disabled. This approach is useful for examining the SMCL producing the special markup effects that are evident when you use `help`. Suppose that you see code you want to emulate in your own help files. Then `viewsource` *cmdname*.`hlp`[1] will show you how that was done, and you do not need to know exactly where the help file is on your machine, except that it must be on your ado-path. A useful template for producing help files is official Stata's `examplehelpfile.hlp`.[1] Again, you do not need to know where it is, or to search for it, as `viewsource` will find it.

---

1. For Stata 10 and later, the extension `.sthlp` is used in place of `.hlp`.

You do not need to be a Stata programmer, or even be interested in the innards of Stata programs or help files, to find `viewsource` useful. Other text files along your ado-path, which includes your current directory or folder, may be opened, too. However, do-files and log files are most likely to be in the current directory or folder, so just using `view` is more direct. Say that you want to repeat a successful but complicated `graph` command, which you carefully stored in a log file or do-file. Use `view` or `viewsource` and then search inside the Viewer using keywords to locate the command before copying it.

Note the emphasis on viewing. The Viewer is not an editor, so making changes to the file is not possible. Use the Viewer when you are clear that neither you nor others should be changing file content, even if the person in question has the requisite file permissions. That situation certainly applies to StataCorp-produced files. With colleagues and students who could do no end of damage if unchecked, this feature is invaluable. It is a limitation if you really do want to edit the file, but then you should already be thinking how to clone `viewsource` so that it fires up your favorite text editor (or Stata's own Do-file Editor). Variants on this idea already exist in Stataland, but writing your own editing command is a good early exercise for any budding Stata programmer.

The Stata Journal (2006)
**6**, Number 2, pp. 279–280

# Stata tip 31: Scalar or variable? The problem of ambiguous names

Gueorgui I. Kolev
Universitat Pompeu Fabra
Barcelona, Spain
gueorgui.kolev@upf.edu

Stata users often put numeric or string values into scalars, which is as easily done as

```
. scalar answer = 42
. scalar question = "What is the answer?"
```

The main discussion of scalars is in [P] **scalar**. Scalars can be used interactively or in programs and are faster and more accurate than local macros for holding values. This advantage would not matter with a number like 42, but it could easily matter with a number that was very small or very large.

Scalars have one major pitfall. It is documented in [P] **scalar**, but users are often bitten by it, so here is another warning. If a variable and a scalar have the same name, Stata always assumes that you mean the variable, not the scalar. This naming can apply even more strongly than you first guess: recall that variable names can be abbreviated so long as the abbreviation is unambiguous (see [U] **11.2 Abbreviation rules**).

Suppose that you are using the `auto` dataset:

```
. sysuse auto
(1978 Automobile Data)
. scalar p = 0.7
. display p
4099
```

What happened to 0.7? Nothing. It is still there:

```
. display scalar(p)
.7
. scalar list p
       p =              .7
```

What is the 4099 result? The dataset has a variable, `price`, and no other variable names begin with p, so p is understood to mean `price`. Moreover, `display` assumes that if you specify just a variable name, you want to see its value in the first observation, namely, `price[1]`. (The full explanation for that is another story.)

What should users do to be safe?

1. Use a different name. For example, you might introduce a personal convention about uppercase and lowercase. Many Stata users use only lowercase letters (and possibly numeric digits and underscores too) within variable names. Such users

could distinguish scalars by using at least one uppercase letter in their names. You could also use a prefix such as `sc_`, as in `sc_p`. If you forget your convention or if a user of your programs does not know about this convention, Stata cannot sense the interpretation you want. Thus this method is not totally safe.

2. Use the `scalar()` pseudofunction to spell out that you want a scalar. This method is totally safe, but some people find it awkward.

3. Use a temporary name for a scalar, as in

```
. tempname p
. scalar `p' = 0.7
```

Each scalar with a temporary name will be visible while the program in which it occurs is running and only in that program. Temporary names can be used interactively, too. (Your interactive session also counts for this purpose as a program.) In many ways, this method is the best solution, as it ensures that scalars can be seen only locally, which is usually better programming style.

The Stata Journal (2006)
**6**, Number 2, p. 281

# Stata tip 32: Do not stop

Stephen P. Jenkins
Institute for Social and Economic Research
University of Essex
Colchester, UK
stephenj@essex.ac.uk

The `do` command, for executing commands from a file, has one (and only one) option: `nostop`. As the online help file for `do` indicates, the option "allows the do-file to continue executing even if an error occurs. Normally, Stata stops executing the do-file when it detects an error (nonzero return code)."

This option can be useful in a variety of circumstances. For example,

1. You wish to apply, in a do-file, the same set of commands to data referring to different groups of subjects or different periods. The commands for each dataset might fail with an error because, say, there are no relevant observations or an `ml` problem may not converge. Running the do-file with the `nostop` option will allow you to get the desired results for all datasets that do not produce an error and, at the same time, identify the potential source of error in the others. The `nostop` option is most useful in initial analyses. Error sources in particular datasets, once identified, can be trapped by using `capture`; if necessary, alternative action may be taken.

2. You have written a command and want to produce a script certifying that using incorrect syntax leads to an appropriate error. There may be several options and several ways in which syntax may be incorrect. With `nostop`, you can test that each of many incorrectly specified options works as expected, while having the do-file containing the test commands run to completion.

The Stata Journal (2006)
**6**, Number 2, pp. 282–283

# Stata tip 33: Sweet sixteen: Hexadecimal formats and precision problems

Nicholas J. Cox
Department of Geography
Durham University
Durham City, UK
n.j.cox@durham.ac.uk

Computer users generally supply numeric inputs as decimals and expect numerical outputs as decimals. But underneath the mapping from inputs to outputs lies software (such as Stata) and hardware that are really working with binary representations of those decimals. Much ingenuity goes into ensuring that conversions between decimal and binary are invisible to you, but occasionally you may see apparently strange side effects of this fact. This problem is documented in [U] **13.11 Precision and problems therein**, but it still often bites and puzzles Stata users. This tip emphasizes that the special hexadecimal format %21x can be useful in understanding what is happening. The format is also documented, but in just one place, [U] **12.5.1 Numeric formats**. Decimal formats such as %23.18f can also be helpful for investigating precision problems.

Binary representations of numbers, using just the two digits 0 and 1, can be difficult for people to interpret without extra calculations. The great advantage of a hexadecimal format, using base 16 (i.e., $2^4$), is that it is closer to base 10 representations while remaining truthful about what can be held in memory as a representation of a number. It is conventional to use the decimal digits 0–9 and the extra digits a–f when base 16 is used. Thus a represents 10 and f represents 15. Hence, at its simplest, hexadecimal 10 represents decimal 16, hexadecimal 11 represents decimal 17, and so forth. (Think of 11 as $1 \times 16^1 + 1 \times 16^0$, for example.) In practice, we want to hold fractions and, as far as possible, some extremely large and extremely small numbers. The general format of a hexadecimally represented number in Stata is thus $mXp$, to be read as $m \times 2^p$. Thus if you use the format %21x with `display`, you can see examples:

```
. di %21x 1
+1.0000000000000X+000
. di %21x -16
-1.0000000000000X+004
. di %21x 1/16
+1.0000000000000X-004
```

You see that 1, −16, and 1/16 are, respectively, $1 \times 2^0$, $-1 \times 2^4$, and $1 \times 2^{-4}$.

The special format is useful to others besides the numerical analysts mentioned in [U] **12.5.1 Numeric formats**. If you encounter puzzling results, looking at the numbers in question should help clarify what Stata is doing and why it does not match your expectation.

Users get bitten in two main ways. First, they forget that most of the decimal digits .1, .2, .3, .4, .5, .6, .7, .8, and .9 cannot be held exactly. Of these, only .5 (1/2) can possibly be represented exactly by a binary approximation; all the others must be held approximately only—regardless of how many bytes are used. To convince yourself of this, see that, e.g., 42.5 can be held exactly,

```
. di %21x 42.5
+1.5400000000000X+005
. di (1 + 5/16 + 4/256) * 2^5
42.5
```

whereas 42.1 cannot be held exactly,

```
. di %21x 42.1
+1.50cccccccccdX+005
. di %23.18f 42.1
  42.100000000000001421
```

Close, but not exact. Second, users forget that although very large or very small numbers can be held approximately, not all possible numbers can be distinguished, even when those numbers are integers within the limits of the variable type being used.

A common source of misery is trying to hold nine-digit integers in numeric variables. If these are identifiers, holding them as `str9` variables is a good idea, but let us focus on what often happens when users read such integers into numeric variables. This experiment shows the problems that can ensue.

```
. gen pinid = 123456789
. di %9.0f pinid[1]
123456792
. di %21x pinid[1]
+1.d6f3460000000X+01a
```

Stata did not complain, but it did not oblige. The value is off by 3. You will see that the value held is a multiple of 4, as the last two digits 92 are divisible by 4. Did we or Stata do something stupid? Can we fix it?

```
. replace pinid = pinid - 3
(0 real changes made)
```

Trying to subtract 3 gives us the same number, so far as Stata is concerned. What is going on? By default, Stata is using a `float` variable. See [D] **data types** if you want more information. At this size of number, such a variable can hold only multiples of 4 exactly, so we lose many final digits. The remedy, if a numeric variable is needed, is to use a `long` or `double` storage type instead.

The Stata Journal (2006)
**6**, Number 3, pp. 425–427

# Stata tip 34: Tabulation by listing

David A. Harrison
Intensive Care National Audit and Research Centre
London, UK
david.harrison@icnarc.org

The command `list` is often regarded as simply a data management tool for listing observations, but it has several little-used options that make it a useful tool for producing customized tables.

The dataset `auto.dta` contains the 1978 repair records (rated 1–5) for various makes of car. We can use `list` in its typical manner to look at some of the data:

```
. use make rep78 using http://www.stata-press.com/data/r9/auto
(1978 Automobile Data)

. drop if missing(rep78)
(5 observations deleted)

. list in 1/5
```

|     | make          | rep78 |
| --- | ------------- | ----- |
| 1.  | AMC Concord   | 3     |
| 2.  | AMC Pacer     | 3     |
| 3.  | Buick Century | 3     |
| 4.  | Buick Electra | 4     |
| 5.  | Buick LeSabre | 3     |

Suppose that we wish to tabulate the makes of car according to repair record. There is no simple approach to produce tables containing strings; however, if we can modify the data so that the variables in the new dataset represent the columns of our desired table and the observations represent the rows, then we can produce the table with a `list` command. The `reshape` command (see [D] **reshape**) provides the ideal tool to do this:

```
. by rep78 (make), sort: gen row = _n

. reshape wide make, i(row) j(rep78)
(note: j = 1 2 3 4 5)

Data                              long   ->   wide
-----------------------------------------------------------------
Number of obs.                      69   ->      30
Number of variables                  3   ->       6
j variable (5 values)            rep78   ->   (dropped)
xij variables:
                                  make   ->   make1 make2 ... make5
-----------------------------------------------------------------
```

The first command above generated a variable defining the rows of our table. Sorting on `make` will ensure that the makes of car appear in alphabetical order within the table. After this `reshape`, the results of a simple `list` will still not be ideal, as the columns will be headed `make1`, `make2`, etc. We can change this by using the `subvarname` option, which substitutes the characteristic `varname` for each variable as the column heading. Characteristics (see [P] **char**) are named items of text that can be attached to any variable or to the entire dataset. We use `forvalues` to loop through the values of `rep78` creating these characteristics. A previous article in the *Stata Journal* has discussed more complex looping, including using the `levelsof` (previously `levels`) and `foreach` commands to cycle through all values of a variable (Cox 2003). Other options for `list` remove observation numbers, remove the default horizontal lines every five rows, insert dividers between the columns, and make the columns of equal width:

```
. forvalues i = 1/5 {
  2.     char make`i´[varname] "Repair record `i´"
  3. }
. list make1-make3, noobs sep(0) divider nocompress subvarname
```

| Repair record 1 | Repair record 2  | Repair record 3 |
|-----------------|------------------|-----------------|
| Olds Starfire   | Cad. Eldorado    | AMC Concord     |
| Pont. Firebird  | Chev. Monte Carlo| AMC Pacer       |
|                 | Chev. Monza      | Audi Fox        |
|                 | Dodge Diplomat   | Buick Century   |
|                 | Dodge Magnum     | Buick LeSabre   |
|                 | Dodge St. Regis  | Buick Regal     |
|                 | Plym. Volare     | Buick Riviera   |
|                 | Pont. Sunbird    | Buick Skylark   |
|                 |                  | Cad. Deville    |
|                 |                  | Cad. Seville    |
|                 |                  | Chev. Chevette  |
|                 |                  | Chev. Malibu    |
|                 |                  | Chev. Nova      |
|                 |                  | Fiat Strada     |

*(output omitted)*

Only the first three columns are displayed here because of the width of the page, but if you require more columns than can be displayed in your results window (and are logging your output), you can use the `linesize(#)` option to increase the available width.

Two-way tables can be achieved in a similar manner:

```
. use make rep78 foreign using http://www.stata-press.com/data/r9/auto
(1978 Automobile Data)
. drop if missing(rep78)
(5 observations deleted)
. by rep78 foreign (make), sort: gen row = _n
. qui reshape wide make, i(rep78 row) j(foreign)
. gen str1 rep78txt = string(rep78) if row == 1
(50 missing values generated)
. format rep78txt %-1s
. char rep78txt[varname] "Repair record"
. char make0[varname] "Domestic"
. char make1[varname] "Foreign"
```

```
. list rep78txt make0 make1, noobs sepby(rep78) div noc subvar abbrev(13)
```

| Repair record | Domestic | Foreign |
|---|---|---|
| 1 | Olds Starfire<br>Pont. Firebird | |
| 2 | Cad. Eldorado<br>Chev. Monte Carlo<br>Chev. Monza<br>Dodge Diplomat<br>Dodge Magnum<br>Dodge St. Regis<br>Plym. Volare<br>Pont. Sunbird | |
| 3 | AMC Concord<br>AMC Pacer<br>Buick Century<br>Buick LeSabre<br>Buick Regal | Audi Fox<br>Fiat Strada<br>Renault Le Car |

(*output omitted*)

Do not underestimate what can be achieved with a simple list!

# Reference

Cox, N. J. 2003. Speaking Stata: Problems with lists. *Stata Journal* 3: 185–202.

The Stata Journal (2006)
**6**, Number 3, pp. 428–429

# Stata tip 35: Detecting whether data have changed

William Gould
StataCorp
College Station, TX
wgould@stata.com

Included in the 17 May 2006 update is a new command that you may find useful, `datasignature`. If you have installed the update, type `help datasignature`. If you have not updated, or are unsure, type `update query` to find out, and type `update all` to install.

Here is the result of running `datasignature` on `auto.dta`:

```
. sysuse auto
(1978 automobile data)
. datasig
    74:12(71728):3831085005:1395876116
```

The output is `auto.dta`'s data signature. If you change the data, even just a little bit, the last two numbers will change:

```
. replace mpg = mpg + 1 in 2
(1 real change made)
. datasig
    74:12(71728):1616229321:1400086868
```

If you change the name of a variable, the last two numbers stay the same and the number inside the parentheses changes:

```
. rename mpg miles_per_gallon
. datasig
    74:12(57876):1616229321:1400086868
```

`datasignature` is designed to help those who use data maintained by others and those who worry that they might themselves have accidentally changed their data.

In the latter case, you could save the signature in the dataset,

```
. datasig
    74:12(57876):1616229321:1400086868
. note: `r(datasignature)´
```

and check it later,

```
. notes
_dta:
  1.  from Consumer Reports with permission
  2.  74:12(57876):1616229321:1400086868
```

Another idea is to include **datasignature** at the beginning of logs:

```
───────────────────────────── begin myfile.do ───
log using myfile, replace
sysuse auto, clear
datasig
...
log close
───────────────────────────── end myfile.do ───
```

You can specify a varlist and **if** and **in**, so if you have a large dataset and want to check just part of it, you can write

```
───────────────────────────── begin myfile.do ───
log using myfile, replace
sysuse auto, clear
datasig mpg weight price if foreign
...
log close
───────────────────────────── end myfile.do ───
```

The Stata Journal (2006)
**6**, Number 3, pp. 430–432

# Stata tip 36: Which observations?

Nicholas J. Cox
Department of Geography
Durham University
Durham City, UK
n.j.cox@durham.ac.uk

A common question is how to identify which observations satisfy some specified condition. The easiest answer is often to use `list`, as in

```
. use http://www.stata-press.com/data/r9/auto, clear
(1978 Automobile Data)
. list rep78 if rep78 == 3
  (output omitted)
```

An equivalent is to use `edit` instead. In either case, the basic ingredients to an answer are

1. At least an `if` condition and possibly an `in` condition, too. Even if we start out interested in all observations, the condition of interest will be specified using `if`.

2. The observation numbers themselves. Evidently some commands will show them (`list` and `edit` being examples), but otherwise we will need to work a little harder and do something like

   ```
   . gen long obsno = _n
   ```

   and work with that new variable. Here I spelled out that the variable type to be used is a `long`. Consulting the help for data types shows that an `int` will work for datasets with up to 32,740 observations. The default for a new variable is `float`: this will often be fine, but it is dangerous for very large datasets because not every large integer less than Stata's maximum dataset size can be held exactly.

What other complications will we need to worry about when specifying conditions?

- Precision problems with noninteger values, prominently documented but nevertheless a frequent source of minor grief (e.g., see Cox [2006] and references therein).

- Ties; i.e., more than one observation may satisfy a specified condition.

- Conditions involving string comparisons as well as numeric comparisons.

`list` or `edit` shows us the observation numbers for a particular condition, but not compactly or retrievably. We do not want to have to type out those numbers if we need them for some other purpose. To get a more compact display, one approach uses `levelsof` after generating an observation number variable.

```
. levelsof obsno if rep78 == 3
1 2 4 6 8 9 10 11 13 14 16 19 25 26 27 28 31 32 34 36 37 39 41 42 44 49 50 54 60 65
```

In an (updated) Stata 8, use `levels`, not `levelsof`. The help for `levelsof` shows that you can put the list of observation numbers into a local macro for further manipulation and that this list is accessible immediately after issuing the command as `r(levels)`.

If you want the `obsno` variable for this kind of purpose, you might want it shortly for something similar, so it might as well be left in memory as long as there is plenty to spare. But `obsno` will remain identical in contents to `_n` only as long as the sort order is not changed.

```
. assert obsno == _n
```

is a good way to check whether that remains true. `assert` gives no output if the assertion made is true for every observation, no news thus being good news in this example. See also Gould (2003).

Asking for the levels of an observation number variable works when ties are present and when string comparisons are specified. You can also add whatever other `if` or `in` conditions apply.

The main problem to worry about in practice is the precision problem. Consider

```
. summarize gear
```

| Variable | Obs | Mean | Std. Dev. | Min | Max |
|---|---|---|---|---|---|
| gear_ratio | 74 | 3.014865 | .4562871 | 2.19 | 3.89 |

What if we want to see which observations are equal to the maximum?

```
. levelsof obsno if gear == 3.89
```

shows nothing and so fails to find the observation(s), whereas

```
. levelsof obsno if gear == float(3.89)
56
```

happens to give the right answer, but you will not always be so lucky. In other circumstances, what you see (3.89) might be more rounded than it should be. The best approach in general is to use the saved results produced by commands such as those, which are documented in the manual entry for each command. Thus after `summarize`,

```
. levelsof obsno if gear == r(max)
56
```

gives the right answer, as it does in this example,

```
. levelsof obsno if gear == `r(max)´
56
```

Nevertheless, I recommend using `r(max)` rather than '`r(max)`' because the former gives you access to the maximum precision possible. A similar comment applies to e-class results.

Incidentally, because `levelsof` is r-class it will overwrite the r-class results left behind by `summarize`, so you will need to issue such commands in the right order. Thus if we wanted to see both the maximums and the minimums, we would need to repeat commands. As a variation, we use the `meanonly` option, which despite its name does leave the maximum and minimum in memory.

```
. summarize gear, meanonly
. levelsof obsno if gear == r(max)
56
. summarize gear, meanonly
. levelsof obsno if gear == r(min)
12
```

# References

Cox, N. J. 2006. Stata tip 33: Sweet sixteen: Hexadecimal formats and precision problems. *Stata Journal* 6: 282–283.

Gould, W. 2003. Stata tip 3: How to be assertive. *Stata Journal* 3: 448.

The Stata Journal (2006)
**6**, Number 4, pp. 588–589

# Stata tip 37: **And the last shall be first**

Christopher F. Baum
Department of Economics
Boston College
Chestnut Hill, MA 02467
baum@bc.edu

Mata's built-in function list contains many useful matrix operations, but I recently came upon one that was lacking: the ability to *flip* a matrix along its rows or columns. Either of those operations can readily be done as a Mata statement, but I'd rather not remember the syntax—or have to remember what it is meant to do when I reread the code. So I wrote these two simple functions:[1]

```
mata:
matrix function flipud(matrix X) {
        return(rows(X)>1 ? X[rows(X)..1,.] : X)
}


matrix function fliplr(matrix X) {
        return(cols(X)>1 ? X[.,cols(X)..1] : X)
}
end
```

These functions will flip a matrix ud—upside down (the first row becomes the last)— or lr, left to right (the first column becomes the last). Because the functions take a **matrix** argument, they may be applied to any of Mata's matrix types, including **string** matrices.

Users have asked why one would want to flip a matrix "upside down". As it happens, doing so becomes a handy tool when creating a two-sided linear filter. Say that we have defined a vector x, containing a declining set of weights: a one-sided linear filter. We can turn x into a two-sided set of weights by using flipud():

```
. mata:
                                              ———— mata (type end to exit) ————
: x = (1\0.5\0.25\0.125\0.0625) ; x
            1

  1         1
  2        .5
  3       .25
  4      .125
  5     .0625
```

---

[1]. I thank Mata's principal architect, William Gould, for improvements he suggested that make the code more general.

```
: x = (flipud(x[2..rows(x)]) \ x); x
                1

        1     .0625
        2     .125
        3     .25
        4     .5
        5     1
        6     .5
        7     .25
        8     .125
        9     .0625

: end
```

To decipher that statement, note that `2..rows(x)` refers to the second through last rows of vector `x`. The statement thus flips those rows of `x` upside down and concatenates them to the original `x` by using the *column-join* operator (see [M-2] **op_join**).

As a second example, consider applying both functions to a string matrix:

```
. mata:
                                              ── mata (type end to exit) ──
: Greek2me = ("alpha","beta","gamma"\"delta","epsilon","zeta"\"eta","theta",
> "iota"\"kappa","lambda","mu"\"nu","xi","omicron"\"pi",
> "rho","sigma"\"tau","upsilon","phi"\"chi","psi","omega")
: Greek2me
              1          2          3

        1    alpha       beta       gamma
        2    delta      epsilon     zeta
        3     eta       theta       iota
        4    kappa      lambda       mu
        5     nu         xi        omicron
        6     pi        rho        sigma
        7    tau       upsilon      phi
        8    chi        psi        omega

: lastFirst = fliplr(flipud(Greek2me)); lastFirst
              1          2          3

        1    omega       psi        chi
        2    phi       upsilon      tau
        3   sigma       rho         pi
        4  omicron      xi          nu
        5     mu       lambda      kappa
        6    iota      theta        eta
        7    zeta     epsilon      delta
        8   gamma       beta       alpha

: end
```

The Stata Journal (2006)
**6**, Number 4, pp. 590–592

# Stata tip 38: Testing for groupwise heteroskedasticity

Christopher F. Baum
Department of Economics
Boston College
Chestnut Hill, MA 02467
baum@bc.edu

A natural source of heteroskedasticity in many kinds of data is *group membership*: observations in the sample may be a priori defined as members of groups, and the variance of a series may differ considerably across groups. This concept will also apply to the errors from a linear regression. The assumption of homoskedasticity in the relationship may reasonably hold within each group, but not between groups. This assumption most commonly arises in cross-sectional datasets. In economic data, for instance, the groups may correspond to firms in different industries or workers in different occupations. It could also apply in a time-series context: for instance, the variance of daily temperature may not be constant over the four seasons. In any case, a test for heteroskedasticity of this sort should take this a priori knowledge into account.

How might we test for groupwise heteroskedasticity in a variable or in the errors from a regression? In the context of regression, if we can argue that each group's regression equation satisfies the classical assumptions (including that of homoskedasticity), the $s^2$ computed by `regress` (see [R] **regress**) is a consistent estimate of the group-specific variance of the disturbance process. For two groups, an $F$ test may be constructed, with the larger variance in the numerator; the degrees of freedom are the residual degrees of freedom of each group's regression. Conducting an $F$ test is easy if both groups' residuals are stored in one variable, with a group variable indicating group membership (in this case 1 or 2). The third form of `sdtest` may then be used, with the by(*groupvar*) option, to conduct the $F$ test.

What if there are more than two groups across which we wish to test for equality of disturbance variance, for instance, a set of 10 industries? We may then use the `robvar` command (see [R] **sdtest**), which like `sdtest` expects to find one variable containing each group's residuals, with a group membership variable identifying them. The by(*groupvar*) option is used here as well. The test conducted is that of Levene (1960) labeled as $W_0$, which is robust to nonnormality of the error distribution. Two variants of the test proposed by Brown and Forsythe (1974), which uses more robust estimators of central tendency (e.g., median rather than mean), $W_{50}$ and $W_{10}$, are also computed.

We illustrate groupwise heteroskedasticity with state-level data: 1 observation per year for each of the six states in the New England region of the United States for 1981–2000. We first apply `robvar` to the state-level population series to examine whether the variance of population is constant across states.

```
. use http://www.stata-press.com/data/imeus/NEdata

. robvar pop, by(state)
```

|          |           | Summary of pop |       |
|----------|-----------|----------------|-------|
| state    | Mean      | Std. Dev.      | Freq. |
| CT       | 3276614.5 | 81452.212      | 20    |
| MA       | 6030915.5 | 178354.76      | 20    |
| ME       | 1212718.1 | 46958.538      | 20    |
| NH       | 1094238.9 | 94362.302      | 20    |
| RI       | 1000209.9 | 29548.701      | 20    |
| VT       | 562960.65 | 31310.625      | 20    |
| Total    | 2196276.3 | 1931629.4      | 120   |

```
 W0 =  13.856324   df(5, 114)     Pr > F = 0.00000000

W50 =  11.820938   df(5, 114)     Pr > F = 0.00000000

W10 =  13.306895   df(5, 114)     Pr > F = 0.00000000
```

All forms of the test clearly reject the hypothesis of homoskedasticity across states' population series: hardly surprising when the standard deviation of Massachusetts' (MA) population is six times that of Rhode Island (RI).

We now fit a linear trend model to state disposable personal income per capita, dpipc, by regressing that variable on `year`. The residuals are tested for equality of variances across states with `robvar`.

```
. regress dpipc  year
```

| Source   | SS         | df  | MS         |
|----------|------------|-----|------------|
| Model    | 3009.33617 | 1   | 3009.33617 |
| Residual | 806.737449 | 118 | 6.83675804 |
| Total    | 3816.07362 | 119 | 32.0678456 |

```
Number of obs =     120
F( 1,    118) =  440.17
Prob > F      =  0.0000
R-squared     =  0.7886
Adj R-squared =  0.7868
Root MSE      =  2.6147
```

| dpipc | Coef.     | Std. Err. | t      | P>\|t\| | [95% Conf. | Interval] |
|-------|-----------|-----------|--------|---------|------------|-----------|
| year  | .8684582  | .0413941  | 20.98  | 0.000   | .7864865   | .9504298  |
| _cons | -1710.508 | 82.39534  | -20.76 | 0.000   | -1873.673  | -1547.343 |

```
. predict double eps, residual

. robvar eps, by(state)
```

|          |            | Summary of Residuals |       |
|----------|------------|----------------------|-------|
| state    | Mean       | Std. Dev.            | Freq. |
| CT       | 4.167853   | 1.3596266            | 20    |
| MA       | 1.618796   | .86550138            | 20    |
| ME       | -2.9841056 | .93797625            | 20    |
| NH       | .51033312  | .61139299            | 20    |
| RI       | -.8927223  | .63408722            | 20    |
| VT       | -2.4201543 | .71470977            | 20    |
| Total    | -6.063e-14 | 2.6037101            | 120   |

```
 W0 =  4.3882072   df(5, 114)     Pr > F = 0.00108562

W50 =  3.2989851   df(5, 114)     Pr > F = 0.00806751

W10 =  4.2536245   df(5, 114)     Pr > F = 0.00139064
```

The hypothesis of equality of variances is soundly rejected by all three `robvar` test statistics, with the residuals for Connecticut, Massachusetts, and Maine possessing a standard deviation considerably larger than those of the other three states.

# References

Brown, M. B., and A. B. Forsythe. 1974. Robust tests for the equality of variances. *Journal of the American Statistical Association* 69: 364–367.

Levene, H. 1960. Robust tests for equality of variances. In *Contributions to Probability and Statistics: Essays in Honor of Harold Hotelling*, ed. I. Olkin, S. G. Ghurye, W. Hoeffding, W. G. Madow, and H. B. Mann, 278–292. Menlo Park, CA: Stanford University Press.

The Stata Journal (2006)
**6**, Number 4, pp. 593–595

# Stata tip 39: In a list or out? In a range or out?

Nicholas J. Cox
Department of Geography
Durham University
Durham City, UK
n.j.cox@durham.ac.uk

Two simple but useful functions, `inlist()` and `inrange()`, were added in Stata 7, but users somehow still often overlook them. The manual entry [D] **functions** gives formal statements on definitions and limits. The aim here is to emphasize with examples how natural and helpful these functions can be.

The question answered by `inlist()` is whether a specified argument belongs to a specified list. That answered by `inrange()` is whether a specified argument falls in a specified range. We can ask the converse question, of not belonging to or falling outside a list or range, by simply negating the function. Thus `!inlist()` and `!inrange()` can be read as "not in list" and "not in range".

These functions can reduce your typing, reduce the risk of small errors, and make your Stata code easier to read and maintain. Thus with the `auto` data in memory, consider the choice for the integer-valued variable `rep78` between older ways of getting a simple listing,

```
. list make rep78 if rep78 == 3 | rep78 == 4 | rep78 == 5
. list make rep78 if rep78 >= 3 & rep78 <= 5
. list make rep78 if rep78 > 2 & rep78 < 6
```

and newer ways of getting the same listing,

```
. list make rep78 if inlist(rep78, 3, 4, 5)
. list make rep78 if inrange(rep78, 3, 5)
```

The examples here are typical of a good way to use `inlist()` or `inrange()`: move directly from feeding arguments to each function to using the results of the calculation. If you wanted to keep the results, you could put them into a variable (or a macro). The result of `inlist()` or `inrange()` is either 1 when the value specified is in range or in list and 0 otherwise (and thus never missing). So, if you use a variable to store results, let it be a byte variable for efficiency in storage.

In more detail: so long as none of the arguments $z, a, b$ is missing, $inrange(z, a, b)$ is true whenever $z \geq a$ and $z \leq b$. Thus `inrange(60, 50, 70)` is true (numerically 1) because $60 \geq 50$ and $60 \leq 70$. However, `inrange(60, 70, 50)` is false (0) because 60 is not $\geq 70$ and 60 is not $\leq 50$. Thus the order of $a$ and $b$ is crucial. There are situations when you are not sure in advance about the ordering of arguments, but you can always use devices such as $inrange(z, \min(a, b), \max(a, b))$ (which tests whether one value is between two others).

The definition of `inrange()` is more complicated when any argument is numeric missing. See [D] **functions** for the precise definitions. The most important example is that `inrange(`$z$`,` $a$`,` `.)` is interpreted as $z \geq a$ and $z <$ . ($z$ greater than or equal to $a$, but not missing). This may look like a bug, but it is really a feature. Even experienced users sometimes forget that in Stata numeric missing is regarded as arbitrarily large. Hence, `z >= 42` will be true for all the missing values of `z`, as well as for all values that are greater than or equal to 42. The longstanding workaround when this is not what you want with regard to missing values is to add the extra condition that `z` is not missing, as in `z >= 42 & z < .`, but `inrange(z, 42, .)` is another way to do this.

The definitions that come into play when any argument is missing imply that `inrange()` is not a good tool to use when you want to test for numeric missings (including any comparisons with extended missing values). For that it is better to use `missing()`, `inlist()`, or combined statements using simple inequalities.

`inlist()` and `inrange()` can often be used with the in-built quantities _n and _N specifying, respectively, the current observation number and the current number of observations. Sometimes users wish to specify that a command should apply to an irregular set of observation numbers, and `if inlist(_n,17,42,99,217)` exemplifies how that could be done with a small set (the limit is 255 numbers and is unlikely to bite in sensible practice). A pitfall here is clearly that any sorting of the dataset will often imply that the observations concerned end up in different positions. Thus saving the results of this computation in a byte variable will often be a good idea. This approach is not better general practice than using criteria such as those based on variable values, but there may be occasions when you will want this feature.

Other examples of the same kind arise with longitudinal or panel data. Recently I wanted to identify the first and last values of a response in each panel, and

```
. by panelvar (timevar): gen y_ends = y if inlist(_n, 1, _N)
```

offers a way to do that. Conversely, `!inlist(_n, 1, _N)` identifies all the others. Whether you prefer that `if` condition to the more traditional `if _n == 1 | _n == _N` is admittedly a matter of taste. Using `in` is not an option here because `in` may not be combined with `by:`.

The examples so far are all for numeric arguments. The arguments of either function can be all numeric or all string. Thus given one character, $c$, `inrange("`$c$`", "a", "z")` tests whether $c$ is one of the 26 lowercase letters of the English alphabet; correspondingly, `inrange("`$c$`", "A", "Z")` tests whether $c$ is one of the 26 uppercase letters of the same alphabet. More generally, `inrange("`*string*`", "a", "z")` tests whether *string* begins with a lowercase letter, and correspondingly for the arguments `"A"`, `"Z"` and uppercase letters. Because lowercase and uppercase letters are typically not adjacent in your computer's character sets, be careful when working with both. If you were indifferent about the distinction between uppercase and lowercase, you could work with `lower("`*string*`)"` or `upper("`*string*`")`.

These examples lead directly to a way of filtering a string variable to select characters
that you want or ignore characters that you don't. Suppose that we wanted to select
only the alphabetic characters in a string variable. Check the variable type to see its
maximum length (18, or whatever), `generate` a new empty-string variable, and then
loop over the characters, adding them to the end of the new variable only if they are as
desired.

```
generate newvar = ""
quietly forvalues i = 1/18 {
        replace newvar = newvar + substr(oldvar,`i´,1) ///
                if inrange(lower(substr(oldvar,`i´,1)),"a","z")
}
```

Commands like this tend to become rather long, but they are not in principle complicated. The attraction of a low-level approach is that you can design exactly the filter
you wish according to the precise problem it is intended to solve.

A further simple but general moral evident from various examples here is that the
power of Stata functions often arises from how they can be combined.

The Stata Journal (2007)
**7**, Number 1, pp. 137–139

# Stata tip 40: Taking care of business[1]

Christopher F. Baum
Department of Economics
Boston College
Chestnut Hill, MA 02467
baum@bc.edu

Daily data are often generated by nondaily processes: for instance, financial markets are closed on weekends and holidays. Stata's time-series date schemes ([U] **11.4.4 Time-series varlists**) allow for daily data, but gaps in time series may be problematic. A model that uses lags or differences will lose several observations every time a gap appears, discarding many of the original data points. Analysis of "business-daily" data often proceeds by assuming that Monday follows Friday, and so on. At the same time, we usually want data to be placed on Stata's time-series calendar so that useful tools such as the `tsline` graph will work and label data points with readable dates; see [TS] **tsline**.

At a recent Stata Users Group presentation in Boston, David Drukker spoke to this point. His solution: generate two date variables, one containing the actual calendar dates, another numbering successive available observations consecutively. The former variable (`caldate`) is `tsset` (see [TS] **tsset**) when the calendar dates are to be used, whereas the latter (`seqdate`) is `tsset` when statistical analyses are to be performed.

We download daily data on the 3-month U.S. Treasury bill rate with Drukker's `freduse` command (Drukker 2006) and retain the August 2005–present data for analysis. (We can also view the data graphically with `tsline`.)

```
. freduse DTB3
(13764 observations read)
. rename daten caldate
. tsset caldate
        time variable:  caldate, 04jan1954 to 05oct2006, but with gaps
. keep if tin(1aug2005,)
(13455 observations deleted)
. label var caldate date
```

These data do not contain observations for weekends and are missing for U.S. holidays. We may not want to drop the observations containing missing data, though, as we may have complete data for other variables: for instance, exchange rate data are available every day. If there were no missing data in our series—only missing observations—we could use Drukker's suggestion and `generate seqdate = _n`. As we have observations for which `DTB3` is missing, we follow a more complex route:

---

1. As of Stata 12, we now have business calendars; see [D] **datetime business calendars**. Also see http://www.youtube.com/watch?v=iiizhsX-100 for a Stata YouTube video on `freduse`.—Ed.

```
. quietly generate byte notmiss = DTB3 < .
. quietly generate seqdate = cond(notmiss, sum(notmiss),.)
. tsset seqdate
        time variable:  seqdate, 1 to 297
```

The variable seqdate is created as the sequential day number for every nonmissing day and is itself missing when DTB3 is missing—allowing us to use this variable in tsset and then use time-series operators (see [U] **11.1.1 varlist**) in generate or estimation commands such as regress. We may want to display the transformed data (or results from estimation, such as predicted values) on a time-series graph. We can just revert to the other tsset:

```
. quietly generate dDTB3 = D.DTB3
. quietly regress dDTB3 L(1/5).dDTB3
. predict double dDTB3hat, xb
(18 missing values generated)
. label var dDTB3 "Daily change in three-month Treasury rate"
. tsset caldate
        time variable:  caldate, 01aug2005 to 05oct2006, but with gaps
. tsline dDTB3 dDTB3hat, yline(0) xlabel(, angle(45))
```

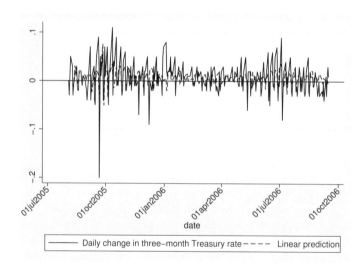

If we retain both the caldate and seqdate variables in our saved dataset, we will always be able to view these data either on a time-series calendar or as a sequential series. In my research, I need to know how many calendar days separate each observed point (1 for Thursday–Friday but 3 for Friday–Monday) and then sum DTB3 by month, weighting each observation by the square root of the days of separation:

```
. tsset seqdate
        time variable:  seqdate, 1 to 297
. quietly generate dcal = D.caldate if seqdate < .
```

```
. quietly generate month = mofd(caldate) if seqdate < .
. format %tm month
. sort month (seqdate)
. quietly by month: generate adjchange = sum(dDTB3/sqrt(dcal))
. quietly by month: generate sumchange = adjchange if _n==_N & month < .
. list month sumchange if sumchange < ., sep(0) noobs
```

| month | sumchange |
|---------|-----------|
| 2005m8 | -.003812 |
| 2005m9 | -.0810769 |
| 2005m10 | .2424316 |
| 2005m11 | -.063453 |
| 2005m12 | .096188 |
| 2006m1 | .2769615 |
| 2006m2 | .099641 |
| 2006m3 | .0142265 |
| 2006m4 | .0938675 |
| 2006m5 | .0350555 |
| 2006m6 | .0327906 |
| 2006m7 | .0304485 |
| 2006m8 | -.083812 |
| 2006m9 | -.123094 |
| 2006m10 | .0442265 |

# Reference

Drukker, D. M. 2006. Importing Federal Reserve economic data. *Stata Journal* 6: 384–386.

The Stata Journal (2007)
**7**, Number 1, p. 140

# Stata tip 41: Monitoring loop iterations

David A. Harrison
Intensive Care National Audit & Research Centre
London, UK
david.harrison@icnarc.org

If, like me, you have ever started a Stata program running and returned hours later to find it still running with no idea of whether it is actually getting anywhere, then you will be looking for a simple method to monitor your loops. The commands `bootstrap` and `jackknife` produce an attractive table of dots for just this purpose by using the undocumented command `_dots`.

Any loop can easily be modified to report its progress by using `_dots`:

```
nois _dots 0, title(Loop running) reps(100)
forvalues i = 1/100 {
    (main body of loop)
    nois _dots `i' 0
}
Loop running (100)
──────┼─── 1 ───┼── 2 ──┼── 3 ──┼── 4 ──┼── 5
..................................................      50
..................................................     100
```

The first `_dots` command, called with the argument 0, sets up the graduated header line. The title and number of repetitions are optional. Further calls to `_dots` take two arguments: the repetition number and a return code. The return code 0 (as used above) indicates a successful repetition, and a dot is displayed. Alternative return codes produce a green 's' ($-1$) or a red 'x' (1), 'e' (2), 'n' (3) or '?' (any other value).

Below is a more complicated example using a `while` loop. Here, the loop runs until 70 successes are achieved. For this contrived example, each iteration succeeds at random with 80% probability. Successes are reported with a dot (.) and failures with an x.

```
nois _dots 0, title(Looping until 70 successes...)
local rep 1
local nsuccess 0
while `nsuccess' < 70 {
    local fail = uniform() < .2
    local nsuccess = `nsuccess' + (`fail' == 0)
    nois _dots `rep++' `fail'
}
Looping until 70 successes...
──────┼─── 1 ───┼── 2 ──┼── 3 ──┼── 4 ──┼── 5
xx...x..x...xx...xx....x..................x.........      50
.x..............x..x.x.x....x.......
```

The Stata Journal (2007)
**7**, Number 1, pp. 141–142

# Stata tip 42: The overlay problem: Offset for clarity

James Cui
Department of Epidemiology and Preventive Medicine
Monash University
Melbourne, Australia
james.cui@med.monash.edu.au

A common graphical problem often arises when one graph axis shows a discrete scale and the other shows a continuous scale. The discrete scale could, for example, represent distinct categories or a series of times at which data were observed. If we want to show several quantities on the continuous axis, matters may easily become confused—and confusing—when some of those quantities are close, especially if they are shown as confidence or other intervals. One answer to this overlap problem is to offset for clarity.

For example, in longitudinal studies, we often need to draw the mean response and 95% confidence intervals of a continuous variable for several categories over the follow-up period. However, the confidence intervals can overlap if the difference between the mean responses is small. Consider an example closely based on one in Rabe-Hesketh and Everitt (2007, 144–166). Mean and standard deviation of depression score, `dep` and `sddep`, have been calculated for each of five visits and two treatment groups, `visit` and `group`. The number of subjects in each combination of visit and group is also given as `n`, so that approximate 95% confidence limits `high` and `low` can be based on twice the standard error, `sddep` $/\sqrt{n}$. See table 1.

Table 1. Mean and standard deviation of depression score over visit

| visit | group | dep | sddep | n | high | low |
|---|---|---|---|---|---|---|
| 1 | Placebo | 16.48 | 5.28 | 27 | 18.51 | 14.45 |
| 1 | Estrogen | 13.37 | 5.56 | 34 | 15.28 | 11.46 |
| 2 | Placebo | 15.89 | 6.12 | 22 | 18.50 | 13.28 |
| 2 | Estrogen | 11.74 | 6.58 | 31 | 14.10 | 9.38 |
| 3 | Placebo | 14.13 | 4.97 | 17 | 16.54 | 11.72 |
| 3 | Estrogen | 9.13 | 5.48 | 29 | 11.17 | 7.09 |
| 4 | Placebo | 12.27 | 5.85 | 17 | 15.11 | 9.43 |
| 4 | Estrogen | 8.83 | 4.67 | 28 | 10.60 | 7.06 |
| 5 | Placebo | 11.40 | 4.44 | 17 | 13.55 | 9.25 |
| 5 | Estrogen | 7.31 | 5.74 | 28 | 9.48 | 5.14 |

To plot these results, we first use `clonevar` to make a copy of `visit` as `x`: that way, `x` inherits format and value labels as well as values from `visit`, not important here but useful in other problems. We copy so that the original `visit` remains unchanged. Adding and subtracting a small value depending on `group` offsets the two groups. Clearly, the value here, 0.05, can be varied according to taste. If there had been three groups, we could have left one where it was and moved the other two. Because the number of groups is either even or odd, a symmetric placement around integer values on the discrete axis can thus be achieved either way.

```
. use depression

. clonevar x = visit

. replace x = cond(group == "Placebo", x - 0.05, x + 0.05)
x was byte now float
(10 real changes made)

. twoway (connected dep x if group == "Placebo", lpattern(solid) msymbol(D))
>        (connected dep x if group == "Estrogen", lpattern(dash) msymbol(S))
>        (rcap high low x if group == "Placebo")
>        (rcap high low x if group == "Estrogen")
>        , xlabel(1 2 3 4 5) ylab(5(5)20, format(%5.0f))
>        xtitle("Visit") ytitle("Depression score")
>        legend(pos(1) ring(0) col(1) order(1 "Placebo" 2 "Estrogen"))
```

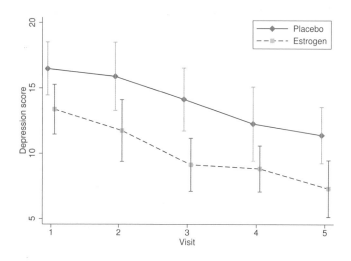

Figure 1. Mean depression score and 95% confidence intervals over visit

# Reference

Rabe-Hesketh, S., and B. Everitt. 2007. *A Handbook of Statistical Analyses Using Stata*. 4th ed. Boca Raton, FL: Chapman & Hall/CRC.

The Stata Journal (2007)
**7**, Number 1, pp. 143–145

# Stata tip 43: Remainders, selections, sequences, extractions: Uses of the modulus

Nicholas J. Cox
Department of Geography
Durham University
Durham City, UK
n.j.cox@durham.ac.uk

The `mod(x, y)` function produces remainders from division. It yields the remainder or residue when `x` is divided by `y`. The manual or online help definition is that `mod(x, y)` yields the modulus of `x` with respect to `y`. Mathematically, this definition is an abuse of terminology, but one that Stata shares with many other computing languages. In mathematics, the modulus is the divisor; somehow a few decades back in computing the term was transferred to the remainder.

Like several other functions, `mod()` may at first seem fairly trivial, so here are examples of some of its uses. All illustrations will be for first arguments (dividends) that are zero or positive integers and second arguments (divisors) that are positive integers. Stata's definition is more general, and yet more general definitions are possible, but the illustrations will show the main idea and cover most practical applications. Texts on discrete mathematics or the mathematics behind computing give fuller treatments (Biggs 2002; Knuth 1997; Graham, Knuth, and Patashnik 1994), but we need none of that material here. Authors often discuss these ideas under the heading of congruences.

How should you play with functions like `mod()` to get to know them? First, there is `display`:

```
. display mod(1,2)
1
. display mod(2,2)
0
. display mod(3,2)
1
```

One useful device is a loop to get several results at once:

```
. forvalues i = 0/8 {
.     display "`i´ " mod(`i´, 3)
. }
```

Second, there is `generate`, typically followed by `list`:

```
. set obs 9
. generate mod3 = mod(_n - 1, 3)
. list mod3
```

You can use the observation numbers _n, which are integers 1 and up, to produce variables corresponding to successive integers.

Third, there is Mata, released in Stata 9:

```
. mata
: x = (0..8)
: mod(x, 3)
```

The first illustration, dividing 1, 2, and 3 by 2, points up a useful detail. Evidently, on division by 2, odd numbers have remainder 1 and even numbers, remainder 0. This example gives a way of characterizing odd and even in Stata. Suppose that you want to specify every other observation. Then

```
if mod(_n, 2) == 1
```

specifies odd-numbered observations and

```
if mod(_n, 2) == 0
```

specifies even-numbered observations. No new variable need be created, as Stata does the necessary calculations on the fly. We can be even more concise:

```
if mod(_n, 2)
```

selects odd observation numbers. Given mod(_n, 2), Stata evaluates it as 1 whenever _n is odd, which is nonzero and therefore true. Further,

```
if !mod(_n, 2)
```

selects even, as mod(_n, 2) is 0 whenever _n is even, but that result is flipped to 1 by the operator !, giving again 1, nonzero and true.

The idea extends easily to other divisors; for example, if mod(year, 10) == 0 or if !mod(year, 10) selects values of year divisible by 10 such as 1990 or 2000, and if !mod(year - 5, 10) selects years such as 1995 or 2005 (but not 1990 or 2000).

Now let us turn to sequences. For integers $x$ from 0 up, mod($x$, 3) is

0 1 2 0 1 2 0 1 2 ...

and for any positive integer $y$, mod($x$, $y$) repeats cycles of 0 to $y - 1$. You may often want to add 1 to get, e.g.,

1 2 3 1 2 3 1 2 3 ...

Here you should use in Stata 1 + mod($x$, 3) and in Mata 1 :+ mod($x$, 3)—note the elementwise operator :+.

You could get such sequences in other ways. Using cond() (Kantor and Cox 2005), we could type for observation numbers _n that run 1 upwards cond(mod(_n, 3) == 0, 3, mod(_n, 3)), giving the same result.

Hence, you now have a basic recipe for generating repetitive sequences. You may know that this functionality is wired into egen's seq() function, but the approach from first principles has merit, too.

Extracting digits is yet another application. In the shadow world between numbers and strings dwell numeric identifiers and run-together dates (20070328 for 28 March 2007) or times (112233 for 11:22:33). Whether such beasts are best processed as numbers or strings can be a close call. Conversion functions `real()` and `string()` are available to throw each to the other side of the divide.

Suppose that your beasts arrive as numeric. `mod(112233, 100)` extracts the last two digits. Hence, second arguments that are $10^k$ will extract the last $k$ digits from integers.

Other subsequences of digits require a little more work. We could get the first two, the second two, and the third two digits like this:

```
. local first = floor(112233/10000)
. local second = floor(mod(112233, 10000) / 100)
. local third = mod(112233, 100)
. display `first'`second'`third'
112233
```

For more on `floor()` and its twin `ceil()`, see Cox (2003). You could also use `int()` here. An alternative is to work with (say) `real(substr(string(112233),1,2))`.

Naturally, if what you are given is just 112233, you do not need Stata or even a computer to extract digits. Rather, these are examples of the kind you can try for yourself to see what is necessary to convert information given in variables from one form to another.

# References

Biggs, N. L. 2002. *Discrete Mathematics.* Oxford: Oxford University Press.

Cox, N. J. 2003. Stata tip 2: Building with floors and ceilings. *Stata Journal* 3: 446–447.

Graham, R. L., D. E. Knuth, and O. Patashnik. 1994. *Concrete Mathematics: A Foundation for Computer Science.* Reading, MA: Addison–Wesley.

Kantor, D., and N. J. Cox. 2005. Depending on conditions: A tutorial on the cond() function. *Stata Journal* 5: 413–420.

Knuth, D. E. 1997. *The Art of Computer Programming, Volume 1: Fundamental Algorithms.* Reading, MA: Addison–Wesley.

The Stata Journal (2007)
**7**, Number 2, pp. 266–267

# Stata tip 44: Get a handle on your sample

Ben Jann
ETH Zürich
Zürich, Switzerland
jann@soz.gess.ethz.ch

Researchers producing careful and reproducible statistical analyses need to keep track of precisely which observations are used by commands. Consider `regress` and similar commands as leading examples. The observations used will depend on any `if` or `in` conditions, any weights specified, and the incidence of missing values. Typically, you will want to look at results for `regress` together with those from other commands. For that you want the same observations to be used. Even when there is comparison with results for different subsets, you also need to monitor which observations are used by which commands.

`if` and `in` conditions and the use of weights are explicit in your command syntax, so you have only yourself to blame if you fail to consider their consequences. Stata, however, does not make a great fuss about excluding missing values from your analyses, so more attention is needed to this detail. Since most substantial statistical datasets contain missing values in at least some of the variables, the issue can arise often.

Researchers commonly start with a simple model and then add more predictors or covariates. At each step, some observations may be excluded because values are missing in the extra variables. As long as the proportion of missing values is not too large, you may not care much about them. However, correct interpretation of the results hinges on the subset of observations used remaining identical.

A brute force approach to the problem is to `keep` only those observations being used (or conversely to `drop` the others). But this method can create as many problems as it solves. Any number of different subsets analyzed would mean as many different datasets and consequent awkwardness in setting up comparisons.

A much better way to get a grip on the samples being used is to construct binary indicator or dummy variables that mark the observations used in any analysis. Their values should be 0 for excluded observations and 1 for included observations. With such variables, corresponding `if` conditions may be specified as desired.

Ado-file programmers (see [U] **17 Ado-files**) face a similar problem and have a special command to solve it. If you look at the source code of ado-files (using [P] **viewsource**; Cox 2006), you will often find the command `marksample touse` near the start and many `if 'touse'` qualifiers after. Although `marksample` can be used only within programs, [P] **mark** documents two other "programmer's commands", `mark` and `markout`, that prove to be handy outside ado-files, as I will now show.

Suppose that you are analyzing a dataset containing the variables `x`, `y`, and `z`, all of which contain some missing values; a group variable, `g`; and analytic weights, `w`. The analysis should be restricted to group `g == 1`. To ensure that the same observations are used throughout the analysis, type

```
. mark touse [aw=w] if g == 1
. markout touse x y z
```

at the beginning of your analysis and include `if touse` in all later commands, as in

```
. reg x y [aw=w] if touse
```

The first command, `mark`, generates a marker variable `touse` (read: "to use") that is set to 1 in observations satisfying the `if` qualifier and having a strictly positive, nonmissing weight and is set to 0 in all other observations.

The second command, `markout`, recodes `touse` to 0 if any of the specified variables contains missing. (If your data are `svyset`, you might want to omit the weights from the first command and add a third line reading `svymarkout touse`; see [SVY] **svymarkout**.)

There are other possible ways to generate the marker variable. You could, for example, use the `missing()` function (see [D] **functions**), or you could code

```
. quietly regress x y z [aw=w] if g==1
. generate byte touse = e(sample)
```

However, using `mark` and `markout` is simple and general. Often it is a good idea to `count if touse` and check that the number of observations used remains the same as that given.

If you need to have a one-liner, then define a program such as

```
program marktouse
version 8
syntax anything(id="markvar") [if] [in] [aw fw iw pw]
gettoken markvar varlist : anything
mark `markvar' `if' `in' [`weight'`exp']
markout `markvar' `varlist'
end
```

and use it as in

```
. marktouse touse x y z [pw=w] if g == 1
```

# Reference

Cox, N. J. 2006. Stata tip 30: May the source be with you. *Stata Journal* 6: 149–150.

The Stata Journal (2007)
**7**, Number 2, pp. 268–271

# Stata tip 45: Getting those data into shape[1]

Christopher F. Baum
Department of Economics
Boston College
Chestnut Hill, MA 02467
baum@bc.edu

Nicholas J. Cox
Department of Geography
Durham University
Durham City, UK
n.j.cox@durham.ac.uk

Are your data in shape? That is, are they in the structure that you need to conduct the analysis you have in mind? Data sources often provide the data in a structure that is suitable for presentation but clumsy for statistical analysis. One of the key data management tools that Stata provides is `reshape`; see [D] **reshape**. If you need to modify the structure of your data, you should be familiar with `reshape` and its two functions: `reshape wide` and `reshape long`. In this tip, we discuss how two applications of `reshape` may be the solution to some knotty data management problems.

As a first example, consider this question posted on Statalist by an individual who has a dataset in the wide form:

| country | tradeflow | Yr1990 | Yr1991 |
|---------|-----------|--------|--------|
| Armenia | imports | 105 | 120 |
| Armenia | exports | 90 | 100 |
| Bolivia | imports | 200 | 230 |
| Bolivia | exports | 80 | 115 |
| Colombia | imports | 100 | 105 |
| Colombia | exports | 70 | 71 |

He would like to reshape the data into long form:

| country | year | imports | exports |
|---------|------|---------|---------|
| Armenia | 1990 | 105 | 90 |
| Armenia | 1991 | 120 | 100 |
| Bolivia | 1990 | 200 | 80 |
| Bolivia | 1991 | 230 | 115 |
| Colombia | 1990 | 100 | 70 |
| Colombia | 1991 | 105 | 71 |

---

1. This tip was updated to use the new command `import delimited` rather than `insheet`.—Ed.

We must exchange the roles of years and tradeflows in the original data to arrive at the desired structure, suitable for analysis as `xt` data. This exchange can be handled by two successive applications of `reshape`:

```
. reshape long Yr, i(country tradeflow)
(note: j = 1990 1991)
Data                              wide   ->   long

Number of obs.                       6   ->       12
Number of variables                  4   ->        4
j variable (2 values)                    ->   _j
xij variables:
                          Yr1990 Yr1991   ->   Yr
```

This transformation swings the data into long form with each observation identified by `country`, `tradeflow`, and the new variable `_j`, taking on the values of year. We now perform `reshape wide` to make imports and exports into separate variables:

```
. rename _j year
. reshape wide Yr, i(country year) j(tradeflow) string
(note: j = exports imports)
Data                              long   ->   wide

Number of obs.                      12   ->        6
Number of variables                  4   ->        4
j variable (2 values)         tradeflow  ->   (dropped)
xij variables:
                                    Yr   ->   Yrexports Yrimports
```

If we transform the data to wide form once again, the `i()` option contains `country` and `year`, as those are the desired identifiers on each observation of the target dataset. We specify that `tradeflow` is the `j()` variable for `reshape`, indicating that it is a `string` variable. The data now have the desired structure. Although we have illustrated this double-reshape transformation with only a few countries, years, and variables, the technique generalizes to any number of each.

As a second example of successive applications of `reshape`, consider the World Bank's World Development Indicators (WDI) dataset.[2] Their extract program generates a comma-separated value (CSV) database extract, readable by Excel or Stata, but the structure of those data hinders analysis as panel data. For a recent year, the header line of the CSV file is

```
"Series code","Country Code","Country Name","1960","1961","1962","1963",
"1964","1965","1966","1967","1968","1969","1970","1971","1972","1973",
"1974","1975","1976","1977","1978","1979","1980","1981","1982","1983",
"1984","1985","1986","1987","1988","1989","1990","1991","1992","1993",
"1994","1995","1996","1997","1998","1999","2000","2001","2002","2003","2004"
```

---

2. See http://econ.worldbank.org.

That is, each row of the CSV file contains a *variable* and *country* combination, with the columns representing the elements of the time series.[3]

Our target dataset structure is that appropriate for panel-data modeling, with the variables as columns and rows labeled by country and year. Two applications of `reshape` will again be needed to reach the target format. We first `import delimited` (see [D] **import delimited**) the data and transform the triliteral country code into a numeric code with the country codes as labels:

```
. import delimited using wdiex
. encode countrycode, generate(cc)
. drop countrycode
```

We then must address that the time-series variables are named `var4–var48`, as the header line provided invalid Stata variable names (numeric values) for those columns. We use `rename` (see [D] **rename**) to change v4 to d1960, v5 to d1961, and so on:

```
forvalues i=4/48 {
        rename v`i´ d`=1956+`i´´
}
```

We now are ready to carry out the first `reshape`. We want to identify the rows of the reshaped dataset by both country code (`cc`) and `seriescode`, the variable name. The `reshape long` will transform a fragment of the WDI dataset containing two series and four countries:

```
. reshape long d, i(cc seriescode) j(year)
(note: j = 1960 1961 1962 1963 1964 1965 1966 1967 1968 1969 1970 1971 1972
> 1973 1974 1975 1976 1977 1978 1979 1980 1981 1982 1983 1984 1985 1986 1987
> 1988 1989 1990 1991 1992 1993 1994 1995 1996 1997 1998 1999 2000 2001 2002
> 2003 2004)
```

| Data | wide | -> | long |
|------|------|-----|------|
| Number of obs. | 7 | -> | 315 |
| Number of variables | 48 | -> | 5 |
| j variable (45 values) | | -> | year |
| xij variables: | | | |
| d1960 d1961 ... d2004 | | -> | d |

---

3. A variation occasionally encountered will resemble this structure, but with periods in reverse chronological order. The solution here can be used to deal with that problem as well.

```
. list in 1/15
```

|      | cc  | seriesc~e | year | countryname | d        |
|------|-----|-----------|------|-------------|----------|
| 1.   | AFG | adjnetsav | 1960 | Afghanistan | .        |
| 2.   | AFG | adjnetsav | 1961 | Afghanistan | .        |
| 3.   | AFG | adjnetsav | 1962 | Afghanistan | .        |
| 4.   | AFG | adjnetsav | 1963 | Afghanistan | .        |
| 5.   | AFG | adjnetsav | 1964 | Afghanistan | .        |
| 6.   | AFG | adjnetsav | 1965 | Afghanistan | .        |
| 7.   | AFG | adjnetsav | 1966 | Afghanistan | .        |
| 8.   | AFG | adjnetsav | 1967 | Afghanistan | .        |
| 9.   | AFG | adjnetsav | 1968 | Afghanistan | .        |
| 10.  | AFG | adjnetsav | 1969 | Afghanistan | .        |
| 11.  | AFG | adjnetsav | 1970 | Afghanistan | -2.97129 |
| 12.  | AFG | adjnetsav | 1971 | Afghanistan | -5.54518 |
| 13.  | AFG | adjnetsav | 1972 | Afghanistan | -2.40726 |
| 14.  | AFG | adjnetsav | 1973 | Afghanistan | -.188281 |
| 15.  | AFG | adjnetsav | 1974 | Afghanistan | 1.39753  |

The rows of the data are now labeled by year, but one problem remains: all variables for a given country are stacked vertically. To unstack the variables and put them in shape for xtreg (see [XT] xtreg), we must carry out a second reshape that spreads the variables across the columns, specifying cc and year as the $i$ variables and seriescode as the $j$ variable. Since that variable has string content, we use the string option.

```
. reshape wide d, i(cc year) j(seriescode) string
(note: j = adjnetsav adjsavC02)
```

| Data                  | long       | -> | wide                |
|-----------------------|------------|----|---------------------|
| Number of obs.        | 315        | -> | 180                 |
| Number of variables   | 5          | -> | 5                   |
| j variable (2 values) | seriescode | -> | (dropped)           |
| xij variables:        |            |    |                     |
|                       | d          | -> | dadjnetsav dadjsavC02 |

```
. order cc countryname

. tsset cc year
        panel variable:  cc (strongly balanced)
        time variable:  year, 1960 to 2004
```

After this transformation, the data are now in shape for xt modeling, tabulation, or graphics.

As illustrated here, the reshape command can transform even the most inconvenient data structure into the structure needed for your research. It may take more than one application of reshape to get there from here, but it can do the job.

The Stata Journal (2007)
**7**, Number 2, pp. 272–274

# Stata tip 46: Step we gaily, on we go

Richard Williams
University of Notre Dame
Notre Dame, IN 46556
richard.a.williams.5@nd.edu

The `nestreg` and `stepwise` prefix commands allow users to estimate sequences of nested models. With `nestreg`, you specify the order in which variables are added to the model. So, for example, a first model might include only demographic characteristics of subjects, a second could add attitudinal measures, and a third could add interaction terms. Conversely, with `stepwise`, the order in which variables enter the model is determined empirically. With forward selection, the variable or block of variables that most improves fit will be entered first, followed by the variable or variables that most improve fit given the variables already in the model, and so forth. Variables that do not meet some specified level of significance will never enter the model.

Despite their similarities, the two commands differ dramatically in the amount of detail that they provide. `stepwise` gives the estimates for the final model it fits but tells little about the intermediate models other than the order in which variables were entered. `nestreg`, on the other hand, offers a wealth of information. The results from each intermediate model can be printed and their estimates stored for later use. Particularly useful is that `nestreg` offers several measures of contribution to model fit. Wald statistics, likelihood-ratio chi-squareds, $R^2$ and change-in-$R^2$ statistics, and Bayesian information criterion (BIC) and Akaike information criterion (AIC) measures are available for each intermediate model. Such measures provide a variety of ways of assessing the importance and effect of each variable or set of variables added to the model.

When forward selection is used, there is a relatively easy way to make the results from `stepwise` as informative and detailed as those provided by `nestreg`. Simply fit the models with `stepwise`, and then refit the models with `nestreg`, listing variables in the order they were added by `stepwise`. For example,

```
. sysuse auto
(1978 Automobile Data)

. stepwise, pe(.05): regress price mpg weight length foreign
                     begin with empty model
p = 0.0000 <  0.0500  adding  weight
p = 0.0000 <  0.0500  adding  foreign
p = 0.0069 <  0.0500  adding  length
```

| Source | SS | df | MS | | Number of obs = | 74 |
|---|---|---|---|---|---|---|
| | | | | | F( 3,   70) = | 28.39 |
| Model | 348565467 | 3 | 116188489 | | Prob > F      = | 0.0000 |
| Residual | 286499930 | 70 | 4092856.14 | | R-squared     = | 0.5489 |
| | | | | | Adj R-squared = | 0.5295 |
| Total | 635065396 | 73 | 8699525.97 | | Root MSE      = | 2023.1 |

| price | Coef. | Std. Err. | t | P>\|t\| | [95% Conf. Interval] | |
|---|---|---|---|---|---|---|
| weight | 5.774712 | .9594168 | 6.02 | 0.000 | 3.861215 | 7.688208 |
| foreign | 3573.092 | 639.328 | 5.59 | 0.000 | 2297.992 | 4848.191 |
| length | -91.37083 | 32.82833 | -2.78 | 0.007 | -156.8449 | -25.89679 |
| _cons | 4838.021 | 3742.01 | 1.29 | 0.200 | -2625.183 | 12301.22 |

```
. nestreg, quietly store(m): regress price weight foreign length if e(sample)
Block  1: weight
Block  2: foreign
Block  3: length
```

| Block | F | Block df | Residual df | Pr > F | R2 | Change in R2 |
|---|---|---|---|---|---|---|
| 1 | 29.42 | 1 | 72 | 0.0000 | 0.2901 | |
| 2 | 29.59 | 1 | 71 | 0.0000 | 0.4989 | 0.2088 |
| 3 | 7.75 | 1 | 70 | 0.0069 | 0.5489 | 0.0499 |

Specifying the `quietly` option omitted the estimates from the intermediate models from the output while still showing the various model change statistics. You may wish to omit `quietly` if, for example, changes in coefficients across models as additional variables are added are of interest. The `store(m)` option stored the estimates from the three models as m1, m2, and m3. Storing the results can be useful if we want to replay the results, format the output with some other program, or do more comparisons across models, say, model m1 versus model m3. Using `if e(sample)` guarantees that the same observations are being analyzed by both `nestreg` and `stepwise`. With `stepwise`, observations with missing data on any of the variables specified get excluded from the analysis, even if those variables do not enter the final model. Most critically, the block residual statistics reported by `nestreg` show us how much the addition of each variable increased $R^2$ and the statistical significance of that change. This approach provides a much more tangible feel for the importance and contribution of each variable than does `stepwise` alone.

If we would also like to see likelihood-ratio contrasts between models, as well as the BIC and AIC statistics for each intermediate model, just add the `lr` option:

```
. nestreg, quietly lr: regress price weight foreign length if e(sample)
Block  1: weight
Block  2: foreign
Block  3: length
```

| Block | LL | LR | df | Pr > LR | AIC | BIC |
|-------|-----------|-------|----|---------|----------|----------|
| 1 | -683.0354 | 25.35 | 1 | 0.0000 | 1370.071 | 1374.679 |
| 2 | -670.1448 | 25.78 | 1 | 0.0000 | 1346.29 | 1353.202 |
| 3 | -666.2613 | 7.77 | 1 | 0.0053 | 1340.523 | 1349.739 |

Naturally, keep the usual cautions concerning stepwise procedures in mind. For example, because multiple tests are being conducted, the reported $p$-values are inaccurate. The researcher may therefore wish to use a more stringent significance level for variable entry, e.g., .01, or use a Bonferroni or other adjustment. Chance alone could cause some variables to enter the model, and a different sample might produce a different final model. Forward and backward selection procedures can also result in different final models. But if you are clear that stepwise selection is appropriate and is being conducted correctly, then combining `stepwise` and `nestreg` should be helpful.

The Stata Journal (2007)
**7**, Number 2, pp. 275–279

# Stata tip 47: Quantile–quantile plots without programming

Nicholas J. Cox
Durham University
Durham City, UK
n.j.cox@durham.ac.uk

Quantile–quantile (Q–Q) plots are one of the staples of statistical graphics. Wilk and Gnanadesikan (1968) gave a detailed and stimulating review that still merits close reading. Cleveland (1993, 1994) gave more recent introductions. Here I look at their use for examining fit to distributions. The quantiles observed for a variable, which are just the data ordered from smallest to largest, may be plotted against the corresponding quantiles from some theoretical distribution. A good fit would yield a simple linear pattern. Marked deviations from linearity may indicate characteristics such as skewness, tail weight, multimodality, granularity, or outliers that do not match those of the theoretical distribution. Many consider such plots more informative than individual figures of merit or hypothesis tests and feature them prominently in intermediate or advanced surveys (e.g., Rice 2007; Davison 2003).

Official Stata includes commands for plots of observed versus expected quantiles for the normal (`qnorm`) and chi-squared (`qchi`) distributions. User-written commands can be found for other distributions. You might guess that such graphics depend on the provision of dedicated programs, but much can be done interactively just by combining some basic commands. Indeed, you can easily experiment with variations on the standard plots not yet provided in any Stata program.

Statistical and Stata tradition dictate that we start with the normal distribution and the `auto` dataset. In a departure from tradition, generate `gpm` (gallons per 100 miles) as a reciprocal of `mpg` (miles per gallon) scaled to convenient units and examine its fit to normality. You can calculate the ranks and sample size by using `egen`:

```
. use http://www.stata-press.com/data/r13/auto
(1978 Automobile Data)
. gen gpm = 100 / mpg
. label var gpm "gallons / 100 miles"
. egen rank = rank(gpm)
. egen n = count(gpm)
```

These `egen` functions handle any missing values automatically and can easily be combined with any extra `if` and `in` conditions. You may like to specify the `unique` option with `rank()` if you have many ties on your variable. If you want to fit separate distributions to distinct groups, apply the `by:` prefix, say,

```
. by foreign, sort: egen rank = rank(gpm)
. by foreign: egen n = count(gpm)
```

Next choose a formula for plotting positions given rank $i$ and count $n$. These positions are cumulative probabilities associated with the data. The formula $i/n$ would imply that no value could be larger than the largest observed in the sample and would render the normal quantile unplottable for the same extreme. The formula $(i - 1)/n$ would be similarly objectionable at the opposite extreme. Various alternatives have been proposed, typically $(i - a)/(n - 2a + 1)$ for some $a$: you may choose for yourself. qnorm has $i/(n + 1)$ (i.e., $a = 0$) wired in, but let us take $a = 0.5$ to emphasize our freedom. A minimal plot is now within reach using invnormal(), the normal quantile or inverse cumulative distribution function. Figure 1 is our first stab, with separate fits for the two groups of cars.

```
. gen pp = (rank - 0.5) / n

. gen normal = invnormal(pp)

. scatter gpm normal if foreign, ms(oh) ||
> scatter gpm normal if !foreign, ms(S) yla(, ang(h))
> legend(order(1 "Foreign" 2 "Domestic") ring(0) pos(5) col(1))
> xti(standard normal)
```

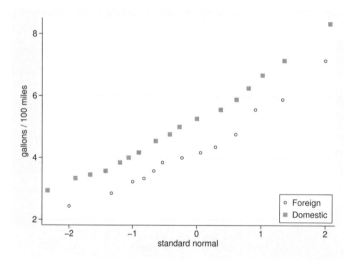

Figure 1. Normal probability plots for gallons per 100 miles for foreign and domestic cars

You might want to fit means and standard deviations explicitly. The easiest way is once again to use egen:

```
. by foreign: egen mean = mean(gpm)

. by foreign: egen sd = sd(gpm)

. gen normal2 = mean + sd * normal

. scatter gpm normal2, by(foreign, note("") legend(off)) ||
> function equality = x, ra(normal2) yla(, ang(h))
> xti(fitted normal) yti(gallons / 100 miles)
```
    *Graph not shown to save space*

Already with just a few lines we can do something not available with qnorm: plotting two or more groups. We can superimpose, as in figure 1, or juxtapose, as in the last example.

Variants of the basic Q–Q plot are also close at hand. Wilk and Gnanadesikan (1968) suggested some possibilities. As is standard practice in examining model fit, we may subtract the general tilt of the Q–Q plot by looking at the residuals, the differences between observed and expected quantiles. These may be plotted against either the expected quantiles or the plotting positions. The two graphs convey similar information. These difference quantile plots might be called DQ plots for short. DQ plots are in essence more demanding than standard Q–Q plots, as they make discrepancies from expectation more evident. As with residual plots, the reference line is no longer a diagonal line of equality but rather the horizontal line of zero difference or residual. Figure 2 shows the two possibilities mentioned. Although gpm is more nearly Gaussian than mpg, some marked skewness remains. Lowess or other smoothing could be used to identify any systematic structure.

```
. gen residual = gpm - normal2
. scatter residual normal2 if foreign, ms(oh) ||
> scatter residual normal2 if !foreign, ms(S)
> legend(order(1 "Foreign" 2 "Domestic") pos(5) ring(0) col(1))
> yla(, ang(h)) yli(0) xti(fitted normal) saving(graph1)
(file graph1.gph saved)
. scatter residual pp if foreign, ms(oh) ||
> scatter residual pp if !foreign, ms(S)
> legend(order(1 "Foreign" 2 "Domestic") pos(5) ring(0) col(1))
> yla(, ang(h)) yli(0) xti(plotting position) saving(graph2)
(file graph2.gph saved)
. graph combine graph1.gph graph2.gph
```

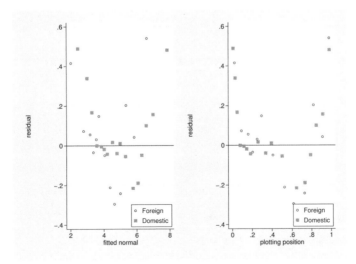

Figure 2. DQ plots for gallons per 100 miles for foreign and domestic cars and normal distribution. Residual versus (left) fitted quantile and (right) plotting position.

The example distribution, the normal, is specified by a location parameter and a scale parameter. This fact gives the flexibility of either fitting parameters or not fitting parameters first. If the theoretical distribution is also specified by one or more shape parameters, we would need to specify those first.

Turning away from the normal, we close with different examples. Q–Q plots and various relatives are prominent in work on the statistics of extremes (e.g., Coles 2001; Reiss and Thomas 2001; Beirlant et al. 2004) and more generally in work with heavy- or fat-tailed distributions. One way of using Q–Q plots is as an initial exploratory device, comparing a distribution, or its more interesting tail, with some reference distribution. For exponential distributions,

```
. generate exponential = -ln(1 - pp)
```

and plot data against that. On such plots, distributions heavier tailed than the exponential will be convex down and those lighter tailed will be convex up (Beirlant et al. 2004). For work with maximums, the Gumbel distribution is a basic starting point.

```
. generate Gumbel = -ln(-ln(pp))
```

Figure 3 is a basic Gumbel plot for annual maximum sea levels at Port Pirie in Australia (data for 1923–1987 from Coles 2001).

```
. scatter level Gumbel, yla(, ang(h)) xti(standard Gumbel)
```

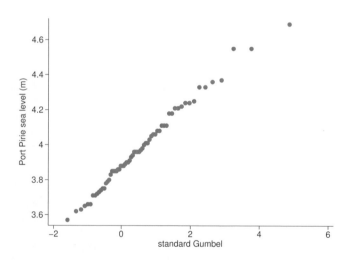

Figure 3. Basic Gumbel plot for annual maximum sea levels at Port Pirie in Australia

The generally good linearity encourages a more formal fit. Convex or concave curves would have pointed to fitting other members of the generalized extreme value distribution family.

# References

Beirlant, J., Y. Goegebeur, J. Segers, and J. Teugels. 2004. *Statistics of Extremes: Theory and Applications.* New York: Wiley.

Cleveland, W. S. 1993. *Visualizing Data.* Summit, NJ: Hobart.

———. 1994. *The Elements of Graphing Data.* Rev. ed. Summit, NJ: Hobart.

Coles, S. 2001. *An Introduction to Statistical Modeling of Extreme Values.* London: Springer.

Davison, A. C. 2003. *Statistical Models.* Cambridge: Cambridge University Press.

Reiss, R.-D., and M. Thomas. 2001. *Statistical Analysis of Extreme Values with Applications to Insurance, Finance, Hydrology, and Other Fields.* Basel: Birkhäuser.

Rice, J. A. 2007. *Mathematical Statistics and Data Analysis.* 3rd ed. Belmont, CA: Duxbury.

Wilk, M. B., and R. Gnanadesikan. 1968. Probability plotting methods for the analysis of data. *Biometrika* 55: 1–17.

# Stata tip 48: Discrete uses for uniform()[1]

Maarten L. Buis
Department of Social Research Methodology
Vrije Universiteit Amsterdam
Amsterdam, The Netherlands
m.buis@fsw.vu.nl

The `uniform()` function produces random draws from a uniform distribution between 0 and 1 ([D] **functions**). `uniform()` is an unusual function. It takes no arguments, although the parentheses are essential to distinguish it from a variable name, and it returns different values each time it is invoked—as many as are needed. Thus, if `uniform()` is used to generate a variable, a different value is created in each observation. This Stata tip focuses on one of its many uses: creating random draws from a discrete distribution where each possible value has a known probability.

A uniform distribution between 0 and 1 means that each number between 0 and 1 is equally likely. So the probability that a random draw from a uniform distribution has a value less than 0.5 is 50%, the probability that such a random draw has a value less than 0.6 is 60%, and so on. The first example below shows how this fact can be used to create a random variable, where the probability of drawing 1 is 60% and that of drawing 0 is 40%. (Kantor and Cox [2005] give a tutorial on the `cond()` function.)

```
gen draw = cond(uniform() < .6, 1, 0)
```

The same result can be achieved even more concisely, given that in Stata a true condition is evaluated as 1 and a false condition as 0 (Cox 2005). `uniform() < .6` is true, and thus evaluated as 1, whenever the function produces a random number less than .6 and is false, and thus evaluated as 0, whenever that is not so.

```
gen draw = uniform() < .6
```

The probability need not be constant. Suppose that the probability of drawing 1 depends on a variable x. We can simulate data for a logistic regression with a constant of $-1$ and an effect of x of 1. In this example, the variable x contains draws from a standard normal distribution.

```
gen x = invnorm(uniform())
gen draw = uniform() < invlogit(-1 + x)
```

Nor is this method limited to random variables with only two values. Consider a distribution in which 1 has probability 30%, 2 probability 45%, and 3 probability 25%.

```
gen rand = uniform()
gen draw = cond(rand < .3,  1, cond(rand < .75, 2, 3))
```

---

1. As of Stata 10.1, the function `uniform()` has been deprecated in favor of `runiform()`. Random draws from a normal distribution are available via the function `rnormal()`. See [D] **functions** or `help random number functions`.—Ed.

A special case of this distribution is one for k integers, say, 1 to k, in which each value is equally likely. Suppose that we simulate throwing a six-sided die, so that values from 1 to 6 are assumed to have probability 1/6. So we need to map all values up to 1/6 to 1, those up to 2/6 to 2, and so forth. That goal is easily achieved by multiplying by 6 and rounding up, using the ceiling function ceil(). Other applications of ceil() were discussed in Stata tip 2 (Cox 2003).

```
gen draw = ceil(6 * uniform())
```

We can use the same principle to simulate draws from a binomial distribution. Recall that a binomial distribution with parameters n and p is the distribution of the number of "successes" out of n trials when the probability of success in each trial is p. One way of sampling from this distribution is to do just that; i.e., draw n numbers from a uniform distribution, declare each number a success if it is less than p, and then count the number of successes (Devroye 1986, 524). Here Mata is convenient. Its equivalent to uniform(), uniform(r, c), creates an r × c matrix filled with random draws from the uniform distribution. Thus we can create a new variable, draw, containing draws from a binomial(100, .3) distribution:

```
mata:
n = 100
p = .3
draw = J(st_nobs(),1,.)                    // matrix to store results
for(i = 1; i <= rows(draw); i++) {         // loop over observations
    trials = uniform(1, n)                 // create n trials
    successes = trials :< p                // success = 1 failure = 0
    draw[i,1] = rowsum(successes)          // count the successes
}
idx = st_addvar("int", "draw")
st_store(., idx, draw)                     // store the variable
end
```

This code is deliberately spun to make its logit clear. Mata learners and experts alike might enjoy working out how to eliminate the loop and how to use fewer variables, while also pondering the possibility of a problem with memory demand for large datasets.

# References

Cox, N. J. 2003. Stata tip 2: Building with floors and ceilings. *Stata Journal* 3: 446–447.

———. 2005. FAQ: What is true or false in Stata?
http://www.stata.com/support/faqs/data/trueorfalse.html.

Devroye, L. 1986. *Non-Uniform Random Variate Generation*. New York: Springer.

Kantor, D., and N. J. Cox. 2005. Depending on conditions: A tutorial on the cond() function. *Stata Journal* 5: 413–420.

The Stata Journal (2007)
**7**, Number 3, pp. 436–437

# Stata tip 49: Range frame plots

Scott Merryman
Risk Management Agency
Kansas City, MO
scott.merryman@gmail.com

One of Edward Tufte's principles for good graph design is to erase nondata ink (Tufte 2001). The axes on a typical scatterplot will extend beyond the range of the observed data. You can minimize the nondata ink by having the axis lines extend only over the range of the data. This can be accomplished by not having the lines extend through the plot region and the region's margin:

```
. sysuse auto
. twoway scatter mpg price, ylabel(minmax, nogrid) xlabel(minmax)
> yscale(nofextend) xscale(nofextend) plotregion(margin(5 2 5 2))
```

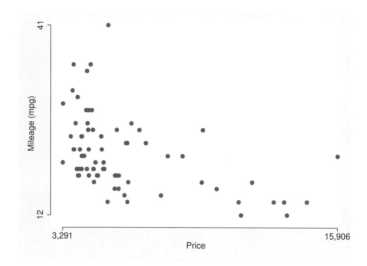

`nofextend` is an undocumented axis-scale option.

To position the origin of the scatterplot at $(0, 10)$, you can adjust the `plotregion(margin())` by the percentage of the observed axis to the total axis length.

```
. sysuse auto

. summarize price, meanonly

. local w = (1 - (r(max) - r(min))/(r(max) - 0))*100

. summarize mpg, meanonly

. local h = (1 - (r(max) - r(min))/(r(max) - 10) )*100

. twoway scatter mpg price, ylabel(minmax, nogrid) xlabel(minmax)
> yscale(nofextend) xscale(nofextend) plotregion(margin(`w´ 2 `h´ 2))
```

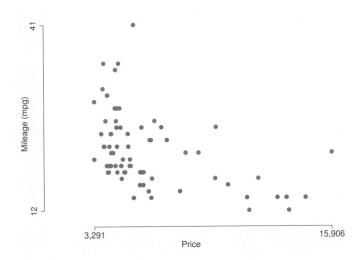

# Reference

Tufte, E. R. 2001. *The Visual Display of Quantitative Information*. 2nd ed. Cheshire, CT: Graphics Press.

The Stata Journal (2007)
**7**, Number 3, pp. 438–439

102

# Stata tip 50: Efficient use of summarize

Nicholas J. Cox
Department of Geography
Durham University
Durham City, UK
n.j.cox@durham.ac.uk

The `summarize` command must be one of the most commonly used Stata commands. Yet, strangely, one of its options is often not used, even though it can be the best solution to a user's problem. Here I flag this neglected `meanonly` option and speculate briefly on why it is often overlooked.

If you fire up `summarize, meanonly`, no results appear in the Results window for you to examine. This lack is deliberate. The option leaves r-class results in memory. (If you are unclear what that means, start with the online help for `return`.) The user must access those results by typing `return list` to see what they are or by feeding one or more results to something else, such as an explicit `display`, `generate`, or `replace` statement. Accessing the saved results should be done promptly after the `summarize, meanonly` command has finished, because results are ephemeral and will not survive beyond the next r-class command that is issued.

The `meanonly` option leaves in memory the mean, as the name implies, in `r(mean)`. However, contrary to what you might guess, it also leaves behind the count of nonmissing values, the sum, the weighted sum, the minimum, and the maximum in appropriately named results. These results are for the last-named variable. Thus, although invoking `summarize, meanonly` with two or more variables is legal, doing so is utterly pointless because results for all but the last will disappear and machine time will be wasted.

Incidentally, if all you want is a count, the `count` command offers a more direct solution; see [D] **count** and Cox (2007).

The difference between `summarize, meanonly` and `summarize` with no options is that the latter also calculates the variance and its square root, the standard deviation. The reason for the `meanonly` option is that this last calculation can be fairly time consuming in large datasets. Thus, if you need to use only one or more of the results left behind after `summarize, meanonly`, then specifying the option will be sensible. Programs or do-files that will be used repeatedly and/or on large datasets are especially suitable. Budding programmers can entertain themselves by identifying StataCorp programs that passed up opportunities for using `summarize, meanonly`. This issue underscores an old joke that you can always speed up a program that was originally written to run slowly.

As a concrete example, one common task is cycling over a set of categories defined by one or more variables. An easy way to do this is to use `egen, group()` to create a variable with integer values 1 and up (and, optionally, value labels with informative text). When you do not know the number of categories in advance,

```
. summarize group, meanonly
```

produces the maximum of `group`, which is the same as the number of categories present. Thus we can feed `r(max)` to whatever code that needs it, possibly a `forvalues` loop.

A small problem remains of explaining why people often overlook this `meanonly` option. I have three guesses. First, `summarize` is one of the commands that Stata users learn early. Typically, it quickly becomes clear that `summarize` does various things and `summarize, detail` does even more. Thus, users tend to feel that they are familiar with the command and do not study its help carefully. Second, the name `meanonly` is in some ways unfortunate and misleading, because much more than the mean is produced. Perhaps a synonym such as `summarize, short` would be a good idea. (Dropping the `meanonly` name is not likely, given the number of programs and commands that would break.) Third, the explanation of `meanonly` in the manual at [R] **summarize** does not give the complete picture on this option.

# Reference

Cox, N. J. 2007. Speaking Stata: Making it count. *Stata Journal* 7: 117–130.

The Stata Journal (2007)
**7**, Number 3, pp. 440–443

# Stata tip 51: Events in intervals

Nicholas J. Cox
Department of Geography
Durham University
Durham City, UK
n.j.cox@durham.ac.uk

Observations in panel or longitudinal datasets are often for irregularly spaced times. Patients may arrive for consultation or treatment, or sites may be visited in the field, at arbitrary times, or other human or natural events may occur with unpredictably uneven gaps. Geophysicists might record earthquakes or political scientists might record incidents of unrest; in either case, events occur with their own irregularity. Such examples could be multiplied. One way researchers seek structure in such data is by counting or summarizing data for each panel over chosen time windows. Usually we look backward: How many times did something happen in the previous 6 months? What was the average of some important variable over observations in the previous 30 days?

To see the precise problem in Stata terms, consider better behaved data in which we have regular observations, say, monthly or daily. Then we can use windows with fixed numbers of observations to calculate the summaries required. `rolling` ([TS] **rolling**) can be especially useful here. This scenario suggests one solution: a dataset with irregular data can be made regular by inserting observations for dates not present in the data. `tsfill` ([TS] **tsfill**) is the key command. In turn the downside of that solution is evident: the bulked-out dataset could be many times larger, even though it carries no extra information.

A more direct solution is possible, typically requiring a few lines of Stata code. Once you grasp the solution, modifying the code for similar problems is easy.

Suppose first that you want to count certain kinds of observations, say, how many times something happened in the previous 30 days. We assume that the data include an identifier (say, `id`) and a daily date (say, **date**) among other variables. A good technique to consider is using the `count` command (Cox 2007a). First, initialize a count variable. Our example will count observations with high blood pressure, so the variable name reflects that:

```
gen n_high_bp = .
```

The idea is to loop over the observations, looking at each one in turn. A basic count will be

```
count if some condition is true &
         observation is in the same panel as this one &
         time is within interval of interest relative to this one
```

The example above is part Stata code, part pseudocode. The parts in *slanted type* are pseudocode. `count` will produce a number in your Results window, but that is less important than `count`'s leaving the result in `r(N)`. We must grab that result before something else overwrites it or it just disappears. We can grab the result and use it:

```
replace n_high_bp = r(N) in this observation
```

We want to repeat this step for each observation. You may know that you can use `forvalues`, often abbreviated `forval`, for automating a loop easily (see Cox 2002 for a tutorial). Suppose that you have 4,567 observations. Then you can type

```
forval i = 1/4567 {
        count if conditions are all satisfied
        replace n_high_bp = r(N) in `i´
}
```

Naturally, your having 4,567 observations is unlikely. So, you could just substitute the correct number for 4,567, or you could think more generally. _N is the number of observations.

```
local N = _N
forval i = 1/`N´ {
        count if conditions are all satisfied
        replace nhighbp = r(N) in `i´
}
```

`forval` is fussy in its syntax, so we cannot use _N directly. The `local` statement sets a local macro, N, to contain its value. Once that exists, we can use its contents by referring to 'N'. As you might guess, i and 'i' refer to another local macro, which the `forvalues` loop brings into being. Each time around the loop it takes on values between 1 and the number of observations.

There are three slots to fill in the pseudocode. Here are three examples to match:

> *some condition is true*
> ```
> inrange(sys_bp, 120, .)
> ```
> *observation is in the same panel as this one*
> ```
> id == id[`i´]
> ```
> *time is within interval of interest relative to this one*
> ```
> inrange(date[`i´] - date, 1, 30)
> ```

Our examples use `inrange()` twice. In the first, we suppose that we are counting how often systolic blood pressure was 120 mm Hg or higher. The `inrange()` condition here has one special and useful feature: it excludes missing values. That is, for example, `inrange(., 120, .)` is 0 (false). I do not expect you to find this behavior intuitive, but it is a feature. In the second, the previous 30 days is specified. For more on `inrange()`, see Cox (2006).

Now we can put it all together.

```
gen n_high_bp = .
local N = _N
quietly forval i = 1/`N' {
    count if inrange(sys_bp, 120, .) & ///
            id == id[`i']           & ///
            inrange(date[`i'] - date, 1, 30)
    replace n_high_bp = r(N) in `i'
}
```

A new detail here is the `quietly` added to the loop to stop a long list of results from being shown. Doing so is not essential. Indeed, at a debugging stage, seeing a stream of output, and being able to check that the results are as desired, is useful and reassuring.

Experienced programmers usually reduce that by one line, starting like this:

```
gen n_high_bp = .
quietly forval i = 1/`= _N' {
```

The shortcut here is documented under the help for `macro`. We are evaluating an expression, here just _N, and using its result, all within the space of the command line.

Stata users sometimes want to do something like this:

```
quietly forval i = 1/`= _N' {
    count if inrange(sys_bp, 120, .) & ///
            id == id[`i']           & ///
            inrange(date[`i'] - date, 1, 30)
    gen n_high_bp = r(N) in `i'
}
```

That code will fail the second time around the loop. The first time around the loop, when i is 1, all will be fine. The new variable **n_high_bp** will be generated. `r(N)` will be put into **n_high_bp[1]**. All the other values of **n_high_bp** will be born as missing. However, the second time around the loop, when i is 2, the `generate` command is illegal, as the **n_high_bp** variable already exists, and you cannot `generate` it again.

The consequence is that within the loop we need to use `replace`. In turn, we need to initialize the variable outside and before the loop (because, conversely, you cannot `replace` something that does not yet exist). Initializing it to missing is good practice, even when we know that the program will overwrite the value in each observation.

There are some disadvantages to this approach. Mainly, it will be a bit slow, especially with large datasets. Having to spell out a few lines of code every time you do something similar could also prove tedious. That task could be an incentive to wrap up the code in a do-file or even a program.

More positively, the logic here should seem straightforward and transparent and fairly easy to modify for similar problems. The key will usually be to pick up whatever we need as a saved result. Suppose that we want to record the mean systolic blood pressure over measurements in the last 30 days. The main change is the use of the `summarize` command rather than the `count` command.

```
gen mean_sys_bp = .
quietly forval i = 1/`= _N´ {
      summarize sys_bp if id == id[`i´] & ///
                          inrange(date[`i´] - date, 1, 30), meanonly
      replace mean_sys_bp = r(mean) in `i´
}
```

For the `meanonly` option of `summarize` and its advantages, see the previous Stata tip (Cox 2007b).

Naturally, there are occasional problems in which the condition that we are considering only observations in the same panel is inappropriate. For those problems, remove or change code like `id == id['i']`.

Finally, the technique is readily adaptable to other kinds of windows, say, with regard to intervals of any predictor or controlling variable.

# References

Cox, N. J. 2002. Speaking Stata: How to face lists with fortitude. *Stata Journal* 2: 202–222.

———. 2006. Stata tip 39: In a list or out? In a range or out? *Stata Journal* 6: 593–595.

———. 2007a. Speaking Stata: Making it count. *Stata Journal* 7: 117–130.

———. 2007b. Stata tip 50: Efficient use of summarize. *Stata Journal* 7: 438–439.

The Stata Journal (2007)
**7**, Number 4, pp. 582–583

108

# Stata tip 52: Generating composite categorical variables

Nicholas J. Cox
Department of Geography
Durham University
Durham City, UK
n.j.cox@durham.ac.uk

If you have two or more categorical variables, you may want to create one composite categorical variable that can take on all the possible joint values. The canonical example for Stata users is given by cross-combinations of `foreign` and `rep78` in the `auto` data. Setting aside missings, `foreign` takes on values of 0 and 1, and `rep78` takes on values of 1, 2, 3, 4, and 5. Hence there are ten possible joint values, which could be 0 and 1, 0 and 2, and so forth. As it happens, only eight occur in the data. If we add the value labels attached to `foreign`, we have Domestic 1, Domestic 2, and so forth.

Writing the values like that raises the question of whether these cross-combinations will be better expressed as string variables or as numeric variables with value labels. On the whole, an integer-valued numeric variable with value labels defined and attached is the best arrangement for any categorical variable, but a string variable may also be convenient, especially if you are producing a kind of composite identifier.

A method often seen is to produce string variables with `tostring` (see [D] **destring**), for example,

```
. tostring foreign rep78, generate(Foreign Rep78)
. gen both = Foreign + Rep78
```

Naturally, there are endless minor variations on this method. A small but useful improvement is to insert a space or other punctuation:

```
. gen both = Foreign + " " + Rep78
```

However, this method is not especially good. `tostring` is really for correcting mistakes, whether attributable to human fault or to some software used before you entered Stata: some variable that should be string is in fact numeric. You need to correct that mistake. `tostring` is a safe way of doing that.

That intended purpose does not stop `tostring` being useful for things for which it was not intended, but there are two specific disadvantages to this method:

1. This method needs two lines, and you can do it in one. That is a little deal.

2. This method could lose information, especially for variables with value labels or with noninteger values. That is, potentially, a big deal.

The second point may suggest using `decode` instead, but my suggestions differ. A better method is to use `egen, group()`. See [D] **egen**.

```
. egen both = group(foreign rep78), label
```

This command produces a new numeric variable, with integer values 1 and above, and value labels defined and attached. Particularly, note the `label` option, which is frequently overlooked.

This method has several advantages:

1. One line.

2. No loss of information. Observations that are identical on the arguments are identical on the results. Value labels are used, not ignored. Distinct noninteger values will also remain distinct.

3. The label is useful—indeed essential—for tables and graphs to make sense.

4. Efficient storage.

5. Extends readily to three or more variables.

Another fairly good method is to use `egen, concat()`.

```
. egen both = concat(foreign rep78), decode p(" ")
```

This command creates a string variable, so it is less efficient for data storage and is less versatile for graphics or modeling. Compared with `tostring`, the advantages are

1. One line.

2. You can mix numeric and string arguments. `concat()` will calculate what is needed.

3. You can use the `decode` option to use value labels on the fly.

4. You can specify punctuation as separator, here a blank.

5. Extends to three or more variables.

The Stata Journal (2007)
7, Number 4, pp. 584–586

# Stata tip 53: Where did my p-values go?

Maarten L. Buis
Department of Social Research Methodology
Vrije Universiteit Amsterdam
Amsterdam, The Netherlands
m.buis@fsw.vu.nl

A useful item in the Stata toolkit is the returned result. For example, after most estimation commands, parameter estimates are stored in a matrix `e(b)`. However, these commands do not return the $t$ statistics, $p$-values, and confidence intervals for those parameter estimates. The aim here is to show how to recover those statistics by using the results that are returned. Consider the following OLS regression:

```
. sysuse auto
(1978 Automobile Data)

. regress price mpg foreign

      Source |       SS       df       MS              Number of obs =      74
-------------+------------------------------           F(  2,    71) =   14.07
       Model |  180261702     2  90130850.8            Prob > F      =  0.0000
    Residual |  454803695    71  6405685.84            R-squared     =  0.2838
-------------+------------------------------           Adj R-squared =  0.2637
       Total |  635065396    73  8699525.97            Root MSE      =  2530.9

-------------+----------------------------------------------------------------
       price |      Coef.   Std. Err.      t    P>|t|     [95% Conf. Interval]
-------------+----------------------------------------------------------------
         mpg |  -294.1955   55.69172    -5.28   0.000    -405.2417   -183.1494
     foreign |   1767.292    700.158     2.52   0.014     371.2169    3163.368
       _cons |   11905.42   1158.634    10.28   0.000     9595.164    14215.67
------------------------------------------------------------------------------
```

# 1   t statistic

The $t$ statistic can be calculated from $t = (\widehat{b} - b)/\mathrm{se}$, where $\widehat{b}$ is the estimated parameter, $b$ is the parameter value under the null hypothesis, and se is the standard error. The null hypothesis is usually that the parameter equals zero; thus we have $t = \widehat{b}/\mathrm{se}$. The $t$ statistic for one parameter (`foreign`) can be calculated by

```
. di _b[foreign]/_se[foreign]
2.5241336
```

All the parameter estimates are also returned in the matrix `e(b)`. A vector of all standard errors is a bit harder to obtain; they are the square roots of the diagonal elements of the matrix `e(V)`. In Mata that vector can be created by typing `diagonal(cholesky(diag(V)))`. Continuing the example, a vector of all $t$ statistics can be computed within Mata by

```
: b = st_matrix("e(b)")´
: V = st_matrix("e(V)")
```

```
: se = diagonal(cholesky(diag(V)))
: b :/ se
                          1

      1  |   -5.282572354
      2  |    2.52413358
      3  |   10.27538518
```

# 2   p-value

The $p$-value can be calculated from $p = 2 * (1 - T(\mathrm{df}, |t|))$, where $T$ is the cumulative distribution function of Student's $t$ distribution, df is the residual degrees of freedom, and $|t|$ is the absolute value of the observed $t$ statistic. The $t$ statistic was calculated before, and the residual degrees of freedom are returned as e(df_r). The absolute value can be calculated by using the abs() function, and $(1 - T(\mathrm{df}, t))$ can be calculated by using the ttail(df, $t$) function. The calculation is put together as follows:

```
. local t = _b[foreign]/_se[foreign]
. di 2*ttail(e(df_r),abs(`t'))
.01383634
```

Using Mata, the vector of all $p$-values is then

```
: df = st_numscalar("e(df_r)")
: t = b :/ se
: 2*ttail(df, abs(t))
                    1

      1  |  1.33307e-06
      2  |  .0138363442
      3  |  1.08513e-15
```

# 3   Confidence interval

The lower and upper bounds of the confidence interval can be calculated as $\widehat{b} \pm t_{\alpha/2}\mathrm{se}$, where $t_{\alpha/2}$ is the critical $t$-value given a significance level $\alpha/2$. This critical value can be calculated by using the invttail(df, $\alpha/2$) function. The lower and upper bounds of the 95% confidence interval for the parameter of foreign are thus given by

```
. di _b[foreign] - invttail(e(df_r),0.025)*_se[foreign]
371.2169
. di _b[foreign] + invttail(e(df_r),0.025)*_se[foreign]
3163.3676
```

The vectors of lower and upper bounds for all parameters follow suit in Mata as

```
: b :- invttail(df,0.025):*se, b :+ invttail(df,0.025):*se
                    1                2

    1    -405.2416661    -183.1494001
    2     371.2169028     3163.367584
    3      9595.1638      14215.66676
```

# 4   Models reporting z statistics

If you are using an estimation command that reports $z$ statistics instead of $t$ statistics, the values become

- _b[foreign]/_se[foreign] for the $z$ statistic;

- 2*normal(-abs('z')) for the $p$-value (where the minus sign comes from the fact normal() starts with the lower tail of the distribution, whereas ttail() starts with the upper tail);

- _b[foreign] - invnormal(0.975)*_se[foreign] for the lower bound of the 95% confidence interval, and _b[foreign] + invnormal(0.975)*_se[foreign] for the upper bound (.975 is used instead of .025 for the same kind of reason).

# 5   Further comments

Often it is unnecessary to do these calculations. In particular, if you are interested in creating custom tables of regression-like output the estimates table command or the tools developed by Jann (2005, 2007) are much more convenient. Similarly, if the aim is to create graphs of regression output, take a good look at the tools developed by Newson (2003) before attempting to use the methods described here. This tip is for situations in which no such command does what you want.

# References

Jann, B. 2005. Making regression tables from stored estimates. *Stata Journal* 5: 288–308.

———. 2007. Making regression tables simplified. *Stata Journal* 7: 227–244.

Newson, R. 2003. Confidence intervals and p-values for delivery to the end user. *Stata Journal* 3: 245–269.

The Stata Journal (2007)
**7**, Number 4, pp. 587–589

# Stata tip 54: Post your results

Philippe Van Kerm
CEPS/INSTEAD
Differdange, Luxembourg
philippe.vankerm@ceps.lu

The command `post` and its companion commands `postfile` and `postclose` are described in [P] **postfile** as "utilities to assist Stata programmers in performing Monte Carlo type experiments". That description understates their usefulness, as `post` is one of the most flexible ways to accumulate results and save them for later use in an external file.

Stata output is displayed in the Results window and can be stored in log files. However, browsing log files and selecting particular results can be tedious and inefficient. Fortunately, there are several alternatives, including the use of `file` (see [P] **file**) or the `estimates` suite of commands (see [R] **estimates**), and `post`, the focus here.

Use of `post` is fully described in [P] **postfile**. The steps are in essence:

1. Call `postfile` to initialize the results file: identify the filename, name its variables, and determine their types.

2. Run the analysis and accumulate the results by repeatedly calling `post`. Each call to `post` adds one observation (record or line) to the results file.

3. Close the results file with `postclose`.

`post` is flexible in what it records: e-class, r-class, or s-class results, string or numeric values, locals, constants, etc. Posted results are recorded without disturbing the data in memory. This is particularly neat: it keeps datasets tidy and allows calling multiple files without interfering with the accumulation of results.

This first example uses the `auto` data. We loop over all possible combinations of `foreign` and `rep78` and save average `price` within each group. Estimates are recorded in a new file named `autoinfo.dta`, which is later opened for displaying results with `tabdisp`.

```
. tempname hdle
. postfile `hdle' foreign rep78 mean using autoinfo
. sysuse auto
(1978 Automobile Data)
. forvalues f=0/1 {
  2.        forvalues r=1/5 {
  3.                summarize price if foreign==`f' & rep78==`r', meanonly
  4.                post `hdle' (`f') (`r') (r(mean))
  5.        }
  6. }
. postclose `hdle'
```

```
. use autoinfo, clear
. label define lf 0 "Domestic car" 1 "Foreign car"
. label values foreign lf
. label variable foreign "Origin of car"
. label variable rep78 "1978 repair record"
. tabdisp rep78 foreign, cell(mean)
```

| 1978<br>repair<br>record | Origin of car<br>Domestic car    Foreign car |
|---|---|
| 1 | 4564.5 |
| 2 | 5967.625 |
| 3 | 6607.074          4828.667 |
| 4 | 5881.556          6261.444 |
| 5 | 4204.5            6292.667 |

This example just shows the technique. In fact, for similar problems, the same effect can be produced easily with statsby (see [D] **statsby**):

```
. sysuse auto
(1978 Automobile Data)
. statsby mean=r(mean), by(foreign rep78) saving(autoinfo2): summarize price
(running summarize on estimation sample)
    (output omitted)
. use autoinfo2
(statsby: summarize)
. tabdisp rep78 foreign, cell(mean)
    (output omitted)
```

However, statsby is too restricted for more elaborate problems. A second example shows computations that store results for each of a series of files, here the numbers of observations and variables. It also demonstrates that graph commands are easily used for displaying results.

```
. tempname hdle
. postfile `hdle´ str20 name str100 label nobs nvar using sysfilesinfo
. sysuse dir
    (output omitted)
. local allfiles "`r(files)´"
. foreach dtafile of local allfiles {
  2.          sysuse `dtafile´, clear
  3.          describe, short
  4.          post `hdle´ ("`dtafile´") (`"`: data label´"´) (r(N)) (r(k))
  5. }
    (output omitted)
. postclose `hdle´
. use sysfilesinfo
. keep if label!=""
(18 observations deleted)
```

```
. replace name = subinstr(name,".dta","",.)
(15 real changes made)
. label variable nobs "Number of observations in dataset"
. label variable nvar "Number of variables in dataset"
. scatter nvar nobs if nobs<250, mlabel(name) mlabposition(12)
```

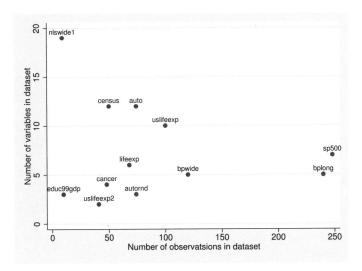

Figure 1. Exploiting posted results

The flexibility of `post` for both collection, when relevant results are posted, and processing, when collected results are analyzed, makes it useful in a broad range of settings, which is different from Monte Carlo simulations.

The Stata Journal (2007)
**7**, Number 4, pp. 590–592

# Stata tip 55: Better axis labeling for time points and time intervals

Nicholas J. Cox
Department of Geography
Durham University
Durham City, UK
n.j.cox@durham.ac.uk

Plots of time-series data show time on one axis, usually the horizontal or $x$ axis. Unless the number of time points is small, axis labels are usually given only for selected times. Users quickly find that Stata's default time axis labels are often not suitable for use in public. In fact, the most suitable labels may not correspond to *any* of the data points. This will arise when it is better to label longer time intervals, rather than any individual times in the dataset.

For example,

```
. webuse turksales
```

reads in 40 quarterly observations for 1990q1 to 1999q4 with a response variable of turkey sales. The default time axis labels with both `line sales t` and `tsline sales` are 1990q1, 1992q3, 1995q1, 1997q3, and 2000q1. These are not good choices for any purpose, even exploration of the data in private.

Label choice is partly a matter of taste, but you might well agree with Stata that labeling every time point would be busy and the result difficult to read. With 40 quarterly values, possible choices include one point per year (10 labels) and one point every other year (5 labels). One possibility is to label every fourth quarter, as that is usually the quarter with highest turkey sales. `summarize` reveals that the times range from 120 to 159 quarters (0 means the first quarter of 1960), so we can type

```
. line sales t, xlabel(123(4)159)
```

Note how we use a *numlist*, 123(4)159, to avoid spelling out every value. The step length is 4 for four quarters. See [U] **11.1.8 numlist** or `help numlist` for more details of *numlist*s. This graph too would need more work before publication, as the labels are still crowded. The text of the labels (e.g., 1990q4) may or may not be judged suitable, depending partly on the readership for the graph.

However, there is another choice: label time intervals (years) and mark the boundaries between those time intervals by ticks. Consider 1990. The four quarters in Stata's units are 120, 121, 122, and 123. Thus we could put text showing the year at a midpoint of 121.5 and ticks showing year boundaries at 119.5 and 123.5. For all years, we should use the *numlist* idea again with the following command to produce figure 1.

```
. line sales t, xtick(119.5(4)159.5, tlength(*1.5))
> xlabel(121.5(4)157.5, noticks format(%tqCY)) xtitle("")
```

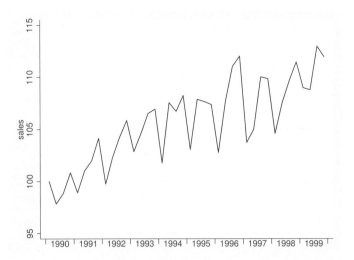

Figure 1. Turkey sales in each quarter. Time axis labels show years (with ticks suppressed) and time axis ticks show year ends.

The most important details here are suppressing the ticks for the axis labels and specifying a format for them. Cosmetic additions include lengthening the ticks compared with the default and suppressing the axis title, which would otherwise be the variable name t (or a variable label if it existed). It is usually clear from the labels what is being shown. Other possibilities include changing the text size for the axis label, changing the angle at which the axis label is shown, and suppressing the century by using a format like %tqY. Those may not be especially attractive, but nevertheless might be forced upon you by practicalities.

The main idea is clearly more general. The axis labels and the axis ticks need not correspond to each other, and it might be good to have fewer labels than ticks for longer series. Monthly and half-yearly data naturally yield to the same method, but use 12 or 2 and not 4 as the step length. Weekly and daily data are more awkward but still manageable.

If you were producing many similar graphs, you might want to automate this process to some degree. The mental arithmetic might easily be more challenging than in the turkey example. Let us imagine daily data for several years. Thus we could put ticks every January 1 and year labels every July 1. That will be adequate precision in practice. Find the first and last years in your data, if necessary by a command like gen year = year(date) followed by summarize. Suppose again that the years are 1990–1999. We can put the needed dates in local macros with a loop:

```
. forvalues y = 1990/1999 {
        local jan `jan' `=mdy(1,1,`y')'
        local jul `jul' `=mdy(7,1,`y')'
}
```

Each time around the loop the daily dates for January 1 and July 1 in each year are calculated on the fly with a call to the `mdy()` function and added to a macro. For more details, see [P] **forvalues** and [P] **macro**, the corresponding help files, or Cox (2002). Once done, the graph command is something like

```
. line whatever t, xlabel(`jul´, format(%tdCY) noticks)
> xtick(`jan´, tlength(*1.5))
```

A key requirement is that the local macros used in the graph command must be visible, by virtue of being in the same interactive session, do-file, or program. That is in essence what `local` means.

Calendar years, meaning here Western calendar years, are clearly not the only possibilities. You could use other boundaries and midpoints for years or other periods defined by other criteria (e.g., academic, financial, fiscal, hydrological, political, religious).

# Reference

Cox, N. J. 2002. Speaking Stata: How to face lists with fortitude. *Stata Journal* 2: 202–222.

The Stata Journal (2008)
**8**, Number 1, pp. 134–136

# Stata tip 56: Writing parameterized text files

Rosa Gini
Regional Agency for Public Health of Tuscany
Florence, Italy
rosa.gini@arsanita.toscana.it

Stata includes several commands for text file manipulation. A good example is the copy command ([D] **copy**). Typing

```
. copy filename1 filename2
```

simply copies `filename1` to `filename2`, regardless of its content.

Often when dealing with text files, you need greater flexibility. Stata can also read and write text files using the `file` suite of commands ([P] **file**). Thereby, you can rewrite a text file by first reading and then writing it with modifications.

```
local filetarget "filename2"
local filesource "filename1"
local appendreplace "replace" /* append or replace */
tempname target source
file open `target´ using `filetarget´, write `appendreplace´ text
file open `source´ using `filesource´, read text
file read `source´ textline
while r(eof) == 0 {
        file write `target´ `"`textline´"´ _n
        file read `source´ textline
}
file close `source´
file close `target´
```

A notable feature of this second way of copying text files is that you can append files to existing files. Even more importantly, you need not copy the file character for character. While rewriting, Stata may substitute the values of local or global macros that have been defined. This allows users to work with a template and produce text with elements substituted for each occasion. Such files may be called "parameterized", as they contain elements constant within a document but variable from document to document.

As an example of the many applications of this simple device, consider the needs of those who periodically access big databases to produce standard reports. The structure of the database is fixed; hence, access will be by a cascade of queries. The queries will be the same every time except for some parameters that change, such as the date (e.g., month, quarter, or year). Stata can access such databases without intermediaries, as SQL code for the queries can be stored in a text file with global macros and be rewritten and executed periodically using the odbc command ([D] **odbc**). A file `query_para.sql` might contain the following simple SQL parameterized code:

```
CREATE TABLE ${year}_AMI AS
SELECT H.ID, H.CODE_PATIENT, H.YEAR, H.SEX, H.AGE
FROM HOSPITALIZATIONS  H, PATHOLOGIES  H_PATHOL
WHERE H.ID=H_PATHOL.ID AND H_PATHOL.DIAGNOSIS="410" AND ${conditions} AND
> H.YEAR=${year} AND H_PATHOL.YEAR=${year};

CREATE INDEX ${year}_ID ON ${year}_AMI (CODE_PATIENT)
TABLESPACE epidemiology;
ANALYZE TABLE ${year}_AMI compute statistics;

CREATE TABLE ${year}_MORTALITY AS
SELECT DISTINCT CASES.CODE_PATIENT, MOR.DEATH_DATE
FROM ${year}_AMI CASES, MORTALITY  MOR
WHERE MOR.CODE_PATIENT=CASES.CODE_PATIENT;
```

This sequence of queries looks for patients hospitalized for AMI (acute myocardial infarction, or heart attack) in a given year and then links the list of patients to the mortality records to obtain data on survival. As the list of patients may be very long, the code computes an index to perform better linkage.

The following code is a template for one Stata session. For example, substitute any connect_options desired for odbc.

```
/* set parameters */
global year = 2005
global conditions "H.AGE>64"
/* rewrite text */
local filetarget "query.sql"
local filesource "query_para.sql"
local appendreplace "replace" /* append or replace */
tempname target source
file open `target' using `filetarget', write `appendreplace' text
file open `source' using `filesource', read text
local i = 1
file read `source' textline
while r(eof) == 0 {
        file write `target' `""`textline'""' _n
        local ++i
        file read `source' textline
}
file close `source'
file close `target'
/* execute queries */
odbc sqlfile("query.sql"), dsn("DataSourceName") [connect_options]
/* load and save generated tables */
foreach table in AMI MORTALITY {
        odbc load table("${year}_`table'"), clear [connect_options]
        save ${year}_`table', replace
}
```

The code will make Stata

1. Write a text file of actual (nonparameterized) SQL code, where ${year} is substituted by 2005 and ${conditions} is substituted by "H.AGE>64".

2. Execute the SQL code via odbc. This may take some time. If an SQL client is available, a practical alternative is to make Stata call that client and ask it

to execute the SQL using the `shell` command ([D] **shell**). This will make the execution of the queries independent of the Stata session.

3. Load the generated tables.

4. Save each of the generated tables as a Stata `.dta` file for later analysis.

The Stata Journal (2008)
**8**, Number 1, pp. 137–138

# Stata tip 57: How to reinstall Stata[1]

Bill Gould
StataCorp
College Station, TX
wgould@stata.com

Sometimes disaster, quite unbidden, may strike your use of Stata. Computers break, drives fail, viruses attack, you or your system administrator do silly—even stupid—things, and Stata stops working or, worse, only partially works, recognizing some commands but not others. What is the solution? You need to reinstall Stata. It will be easier than you may fear.

1. If Stata is still working, get a list of the user-written ado-files you have installed. Bring up your broken Stata and type

   ```
   . ado dir
   ```

   That should list the packages. `ado dir` is a built-in command of Stata, so even if the ado-files are missing, `ado dir` will work. Assuming it does work, let's type the following:

   ```
   . log using installed.log, replace
   . ado dir
   . log close
   ```

   Exit Stata and store the new file `installed.log` in a safe place. Print the file, too. In the worst case, we can use the information listed to reinstall the user-written files.

2. With both your original CD and your paper license handy, take a deep breath and reinstall Stata. If you cannot find your original license codes, call Stata Technical Services.

3. Launch the newly installed Stata. Type `ado dir`. That will either (1) list the files you previously had installed or (2) list nothing. Almost always, the result will be (1).

4. Regardless, `update` your Stata: Type `update query` and follow the instructions.

5. Now you are either done, or, very rarely, you still need to reinstall the user-written files. In that case, look at the original `ado dir` listing we obtained in step 1. One line might read

   ```
   [1] package mf_invtokens from http://fmwww.bc.edu/RePEc/bocode/m
       `MF_INVTOKENS': module (Mata) to convert ...
   ```

---

1. As of Stata 13, Stata is now available via DVDs and download.—Ed.

so you would type

```
. net from http://fmwww.bc.edu/RePEc/bocode/m
. net install mf_invtokens
```

In the case where the package is from http://fmwww.bc.edu, easier than the above is to type

```
. ssc install mf_invtokens
```

Both will do the same thing. `ssc` can be used only to install materials from http://fmwww.bc.edu/. In other cases, type the two `net` commands.

Anyway, work the list starting at the top.

Unless you have had a disk failure, it is exceedingly unlikely that you will lose the user-written programs. If you do not have a backup plan in place for your hard disk, it is a good idea to periodically log the output of `ado dir` and store the output in a safe place.

The Stata Journal (2008)
8, Number 1, pp. 139–141

# Stata tip 58: nl is not just for nonlinear models

Brian P. Poi
StataCorp
College Station, TX
bpoi@stata.com

## 1   Introduction

The `nl` command makes performing nonlinear least-squares estimation almost as easy as performing linear regression. In this tip, three examples are given where `nl` is preferable to `regress`, even when the model is linear in the parameters.

## 2   Transforming independent variables

Using the venerable `auto` dataset, suppose we want to predict the weight of a car based on its fuel economy measured in miles per gallon. We first plot the data:

```
. sysuse auto
. scatter weight mpg
```

Clearly, there is a negative relationship between `weight` and `mpg`, but is that relationship linear? The engineer in each of us believes that the amount of gasoline used to go one mile should be a better predictor of weight than the number of miles a car can go on one gallon of gas, so we should focus on the reciprocal of `mpg`. One way to proceed would be to create a new variable, `gpm`, measuring gallons of gasoline per mile and then to use `regress` to fit a model of `weight` on `gpm`. However, consider using `nl` instead:

```
. nl (weight = {b0} + {b1}/mpg)
(obs = 74)
Iteration 0:  residual SS =  1.19e+07
Iteration 1:  residual SS =  1.19e+07
```

| Source | SS | df | MS |
|--------|------|------|------|
| Model | 32190898.6 | 1 | 32190898.6 |
| Residual | 11903279.8 | 72 | 165323.33 |
| Total | 44094178.4 | 73 | 604029.841 |

|  |  |
|---|---|
| R-squared     = | 0.7300 |
| Adj R-squared = | 0.7263 |
| Root MSE     = | 406.5997 |
| Res. dev.    = | 1097.134 |

| weight | Coef. | Std. Err. | t | P>\|t\| | [95% Conf. Interval] | |
|--------|-------|-----------|---|--------|----------|----------|
| /b0 | 415.1925 | 192.5243 | 2.16 | 0.034 | 31.40241 | 798.9826 |
| /b1 | 51885.27 | 3718.301 | 13.95 | 0.000 | 44472.97 | 59297.56 |

```
Parameter b0 taken as constant term in model & ANOVA table
```

(You can verify that $R^2$ from this model is higher than that from a linear model of `weight` on `mpg`. You can also verify that our results match those from regressing `weight` on `gpm`.)

Here a key advantage of `nl` is that we do not need to create a new variable containing the reciprocal of `mpg`. When doing exploratory data analysis, we might want to consider using the natural log or square root of a variable as a regressor, and using `nl` saves us some typing in these cases. In general, instead of typing

```
. generate sqrtx = sqrt(x)
. regress y sqrtx
```

we can type

```
. nl (y = {b0} + {b1}*sqrt(x))
```

# 3   Marginal effects and elasticities[1]

Using `nl` has other advantages as well. In many applications, we include not just the variable $x$ in our model but also $x^2$. For example, most wage equations express log wages as a function of experience and experience squared. Say we want to fit the model

$$y_i = \alpha + \beta_1 x_i + \beta_2 x_i^2 + \epsilon_i$$

and then determine the elasticity of $y$ with respect to $x$; that is, we want to know the percent by which $y$ will change if $x$ changes by one percent.

Given the interest in an elasticity, the inclination might be to use the `mfx` command with the `eyex` option. We might type

```
. generate xsq = x^2
. regress y x xsq
. mfx compute, eyex
```

These commands will not give us the answer we expect because `regress` and `mfx` have no way of knowing that `xsq` is the square of `x`. Those commands just see two independent variables, and `mfx` will return two "elasticities", one for `x` and one for `xsq`. If $x$ changes by some amount, then clearly $x^2$ will change as well; however, `mfx`, when computing the derivative of the regression function with respect to `x`, holds `xsq` fixed!

The easiest way to proceed is to use `nl` instead of `regress`:

```
. nl (y = {a} + {b1}*x + {b2}*x^2), variables(x)
. mfx compute, eyex
```

---

1. In Stata 11, this example can be done using `regress` with factor-variable notation and the new `margins` command:

```
. regress y x c.x#c.x
. margins, eyex(x) atmeans
```

However, this method only works with polynomials; more general nonlinear functions still require `nl`.—Ed.

Whenever you intend to use `mfx` after `nl`, you must use the `variables()` option. This option causes `nl` to save those variable names among its estimation results.

# 4 Constraints

`nl` makes imposing nonlinear constraints easy. Say you have the linear regression model

$$y_i = \alpha + \beta_1 x_{1i} + \beta_2 x_{2i} + \beta_3 x_{3i} + \epsilon_i$$

and for whatever reason you want to impose the constraint that $\beta_2 \beta_3 = 5$. We cannot use the `constraint` command in conjunction with `regress` because `constraint` only works with linear constraints. `nl`, however, provides an easy way out. Our constraint implies that $\beta_3 = 5/\beta_2$, so we can type

```
. nl (y = {a} + {b1}*x1 + {b2=1}*x2 + (5/{b2})*x3)
```

Here we initialized $\beta_2$ to be 1 because if the product of $\beta_2$ and $\beta_3$ is not 0, then neither of those parameters can be 0, which is the default initial value used by `nl`.

The Stata Journal (2008)
8, Number 1, pp. 142–145

# Stata tip 59: Plotting on any transformed scale

Nicholas J. Cox
Department of Geography
Durham University
Durham City, UK
n.j.cox@durham.ac.uk

Using a transformed scale on one or the other axis of a plot is a standard graphical technique throughout science. The most common example is the use of a logarithmic scale. This possibility is wired into Stata through options `yscale(log)` and `xscale(log)`; see [G-3] *axis_scale_options*. The only small difficulty is that Stata is not especially smart at reading your mind to discern what axis labels you want. When values range over several orders of magnitude, selected powers of 10 are likely to be convenient. When values range over a shorter interval, labels based on multiples of 1 2 5 10, 1 4 7 10, or 1 3 10 may all be good choices.

No other scale receives such special treatment in Stata. However, other transformations such as square roots (especially for counts) or reciprocals (e.g., in chemistry or biochemistry [Cornish-Bowden 2004]) are widely used in various kinds of plots. The aim of this tip is to show that plotting on *any* transformed scale is straightforward. As an example, we focus on logit scales for continuous proportions and percents.

Given proportions $p$, logit $p = \ln\{p/(1 - p)\}$ is perhaps most familiar to many readers as a link function for binary response variables within logit modeling. Such logit modeling is now over 60 years old, but before that lies a century over which so-called logistic curves were used to model growth or decay in demography, ecology, physiology, chemistry, and other fields. Banks (1994), Kingsland (1995), and Cramer (2004) give historical details, many examples, and further references.

The growth of literacy and its complement—the decline of illiteracy—provide substantial examples. In a splendid monograph, Cipolla (1969) gives fascinating historical data but no graphs. Complete illiteracy and complete literacy provide asymptotes to any growth or decay curve, so even without any formal modeling we would broadly expect something like S-shaped or sigmoid curves. Logit scales in particular thus appear natural or at least convenient for plotting literacy data (Sopher 1974, 1979). More generally, plotting on logit scales goes back at least as far as Wilson (1925).

Figure 1 shows how many newly married people could not write their names in various countries during the late nineteenth century, as obtained with data from Cipolla (1969, 121–125) and the following commands:

```
. local yti "% newly married unable to write their names"
. line Italy_females Italy_males France_females France_males Scotland_females
> Scotland_males year, legend(pos(3) col(1) size(*0.8)) xla(1860(10)1900)
> xtitle("") yla(, ang(h)) ytitle(`yti')
```

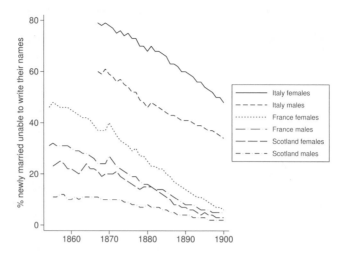

Figure 1. Line plot of illiteracy by sex for various countries in the nineteenth century.

How do we show such data on a logit scale? There are two steps. First, calculate the coordinates you want shown before you graph them. Here we loop over a bunch of variables and apply the transform `logit(`*percent*`/100)`:

```
. foreach v of var *males {
.         gen logit_`v´ = logit(`v´/100)
.         label var logit_`v´ "`: var label `v´´"
. }
```

For tutorials on `foreach` and the machinery used here in looping, see Cox (2002, 2003). Note that, at the same time, we copy variable labels across so that they will show up automatically on later graph legends.

Second, and just slightly more difficult, is to get axis labels as we want them (and axis ticks also, if needed). Even people who work with logits all the time usually do not want to decode that a logit of 0 means 50%, or a logit of 1 means 73.1%, and so forth, even if the `invlogit()` function makes the calculation easy. Logit scales stretch percents near 0 or 100 compared with those near 50. Inspection of figure 1 suggests that 2 5 10(10)80 would be good labels to show for percents within the range of the data. So we want text like 50 to be shown where the graph is showing `logit(50/100)`. The key trick is to pack all the text we want to show and where that text should go into a local macro.

```
. foreach n of num 2 5 10(10)80 {
.         local label `label´ `= logit(`n´/100)´ "`n´"
. }
```

To see what is happening, follow the loop: First time around, local macro 'n' takes on the value 2. `logit(2/100)` is evaluated on the fly (the result is about $-3.8918$) and that is where on our $y$ axis the text "2" should go. Second time around, the same is done for 5 and `logit(5/100)`. And so forth over the numlist 2 5 10(10)80.

Now we can get our graph with logit scale:

```
. line logit_Italy_females logit_Italy_males logit_France_females
> logit_France_males logit_Scotland_females logit_Scotland_males year,
> legend(pos(3) col(1) size(*0.8)) xla(1860(10)1900) xtitle("")
> yla(`label´, ang(h)) ytitle(`yti´)
```

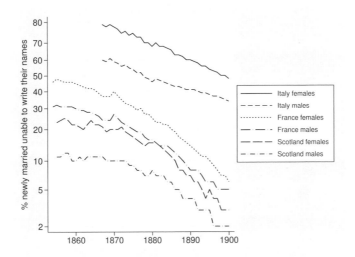

Figure 2. Line plot of illiteracy by sex for various countries in the nineteenth century. Note the logit scale for the response.

Specifically in this example, we now can see data on a more natural scale, complementing the original raw scale. The granularity of the data (rounded to integer percents) is also evident.

Generally, some small details of macro handling deserve flagging. You may be accustomed to a tidy form of local macro definition:

```
. local macname "contents"
```

But the delimiters " " used here would, for this problem, complicate processing given that we do want double quotes inside the macro. Note from the previous `foreach` loop that they can be left off, to advantage.

In practice, you might need to iterate over several possible sets of labels before you get the graph you most like. Repeating the whole of the `foreach` loop would mean that the local macro would continue to accumulate material. Blanking the macro out with

```
. local label
```

will let you start from scratch.

The recipe for ticks is even easier. Suppose we want ticks at 15(10)75, that is, at 15 25 35 45 55 65 75. We just need to be able to tell Stata exactly where to put them:

```
. foreach n of num 15(10)75 {
.         local ticks `ticks' `= logit(`n'/100)'
. }
```

Then specify an option such as `yticks('ticks')` in the graph command.

Finally, note that the local macros you define must be visible to the graph command you issue, namely within the same interactive session, do-file, or program. That is what local means, after all.

In a nutshell: Showing data on any transformed scale is a matter of doing the transformation in advance, after which you need only to fix axis labels or ticks. The latter is best achieved by building graph option arguments in a local macro.

# References

Banks, R. B. 1994. *Growth and Diffusion Phenomena: Mathematical Frameworks and Applications.* Berlin: Springer.

Cipolla, C. M. 1969. *Literacy and Development in the West.* Harmondsworth: Penguin.

Cornish-Bowden, A. 2004. *Fundamentals of Enzyme Kinetics.* 3rd ed. London: Portland Press.

Cox, N. J. 2002. Speaking Stata: How to face lists with fortitude. *Stata Journal* 2: 202–222.

———. 2003. Speaking Stata: Problems with lists. *Stata Journal* 3: 185–202.

Cramer, J. S. 2004. The early origins of the logit model. *Studies in History and Philosophy of Biological and Biomedical Sciences* 35: 613–626.

Kingsland, S. E. 1995. *Modeling Nature: Episodes in the History of Population Ecology.* 2nd ed. Chicago: University of Chicago Press.

Sopher, D. E. 1974. A measure of disparity. *Professional Geographer* 26: 389–392.

———. 1979. Temporal disparity as a measure of change. *Professional Geographer* 31: 377–381.

Wilson, E. B. 1925. The logistic or autocatalytic grid. *Proceedings of the National Academy of Sciences* 11: 451–456.

The Stata Journal (2008)
**8**, Number 2, pp. 290–292

# Stata tip 60: Making fast and easy changes to files with filefilter

Alan R. Riley
StataCorp
College Station, TX
ariley@stata.com

Stata has a command `filefilter` (see [D] **filefilter**) that makes it possible to perform global search-and-replace operations on a file, saving the result to another file. Think of it as a command which copies a file, but in the process of copying it, can search for one text pattern and replace it with another.

As its manual entry points out, `filefilter` has been designed to read and write the input and output files using buffers for speed, and is thus fast at converting even large files. `filefilter` can be used on files which are too large to open in a traditional text editor, and because Stata is programmable, it is possible to use `filefilter` to perform complicated global search-and-replace operations, which would not be possible in most text editors. `filefilter` can even make changes to binary files.

`filefilter` is often used to preprocess files (perhaps to remove invalid characters or to change a delimiter) before reading them into Stata and can be used in many other situations as a useful file-processing tool.

For example, if you have a log file named `x.log` in Windows (where the end-of-line (EOL) character combination is \r\n), and you want to convert the file to have Unix-style EOL characters (\n), you can type in Stata

```
. filefilter x.log y.log, from(\r\n) to(\n)
```

which will replace every occurrence of the Windows EOL character sequence with the Unix EOL character. Equivalently, you could type

```
. filefilter x.log y.log, from(\W) to(\U)
```

because `filefilter` understands \W as a synonym for the Windows EOL character sequence \r\n and \U as a synonym for the Unix EOL character sequence \n.

(For the rest of this tip, I will write \W as the EOL marker, but be sure to use the EOL shorthand in `filefilter` appropriate for your operating system.)

Let's put `filefilter` to use on another example. Imagine that we want to replace all occurrences of multiple blank lines in a file with a single blank line for readability. Changing a file in this way may be desirable after, for example, the command `cleanlog` (Sieswerda 2003) which reads a log file (plain text or SMCL) and removes all command syntax and other extraneous material, leaving behind only output. However, in doing so, `cleanlog` leaves behind multiple adjacent blank lines.

Consider the following lines from a file. (The file below obviously did not result from
`cleanlog`, but it will serve for the purpose of this example.) I will write EOL everywhere
the file contains an end-of-line character sequence.

```
here is a line.  the next two lines are blank in the original file.EOL
EOL
EOL
here is another line.  the next line is blank in the original file.EOL
EOL
this is the last line of the file.EOL
```

The first `filefilter` syntax you might think of would be

```
. filefilter x.log y.log, from(\r\n\r\n) to(\r\n)
```

but it will not do what we want. Because there are EOL characters at the end of
nonblank lines, if all adjacent pairs of EOL characters (\W\W) were replaced with single
EOL characters (\W), the file above would end up looking like

```
here is a line.  the next two lines are blank in the original file.EOL
here is another line.  the next line is blank in the original file.EOL
this is the last line of the file.EOL
```

with no blank lines at all. To have a blank line between sections of output, there must
be two adjacent EOL characters: one at the end of a line, and another on a line all by
itself (the blank line).

Thus, to compress multiple adjacent blank lines down to single blank lines, we need
to replace every occurrence of three adjacent EOL characters with two EOL characters:

```
. filefilter x.log y.log, from(\W\W\W) to(\W\W)
```

We still have not quite achieved the desired result. If we issue the above command
only once, there may still be adjacent empty lines left in the file. We actually need to call
`filefilter` multiple times, each time changing every three newlines to two newlines. I
will assume that `x.log` is the original file, and will use `y.log` and `z.log` as output files
with `filefilter` so that the original file will be left unchanged:

```
. filefilter x.log y.log, from(\W\W\W) to(\W\W)
. filefilter y.log z.log, from(\W\W\W) to(\W\W)
. filefilter z.log y.log, from(\W\W\W) to(\W\W) replace
. filefilter y.log z.log, from(\W\W\W) to(\W\W) replace
. filefilter z.log y.log, from(\W\W\W) to(\W\W) replace
...
```

The above should continue until no more changes are made. We can automate this
by checking the return results from `filefilter` to see if the `from()` pattern was found.
If it was not, we know there were no changes made, and thus, no more changes to be
made:

```
filefilter x.log y.log, from(\W\W\W) to(\W\W)
local nchanges = r(occurrences)
while `nchanges' != 0 {
    filefilter y.log z.log, from(\W\W\W) to(\W\W) replace
    filefilter z.log y.log, from(\W\W\W) to(\W\W) replace
    local nchanges = r(occurrences)
}
...
```

After the code above is executed, `y.log` will contain the desired file, and `z.log` can be discarded. It is possible that the code above will call `filefilter` one more time than is necessary, but unless we have an extremely large file that takes `filefilter` some time to process, we won't even notice.

While it may seem inefficient to use `filefilter` to make multiple passes through a file until the desired result is achieved, it is a fast and easy way to make such modifications. For very large files, Stata's `file` command ([P] **file**) or Mata's I/O functions ([M-4] **io**) could be used to perform such processing in a single pass, but they require a higher level of programming effort.

# Reference

Sieswerda, L. E. 2003. cleanlog: Stata module to clean log files. Statistical Software Components S432401, Department of Economics, Boston College. http://ideas.repec.org/c/boc/bocode/s432401.html.

The Stata Journal (2008)
8, Number 2, pp. 293–294

# Stata tip 61: Decimal commas in results output and data input

Nicholas J. Cox
Department of Geography
Durham University
Durham City, UK
n.j.cox@durham.ac.uk

Given a decimal fraction to evaluate, such as 5/4, Stata by default uses a period (stop) as a decimal separator and shows the result as 1.25. That is, the period separates, and also joins, the integer part 1 and the fractional part 25, meaning here 25/100. Many Stata users, particularly in the United States and several other English-speaking countries, will have learned of such decimal points at an early age and so may think little of this. However, in many other countries, commas are used as decimal separators, so that 1,25 is the preferred way of representing such fractions. This tip is for those users, although it also provides an example of how different user preferences can be accommodated by Stata.

Over several centuries, mathematicians and others have been using decimal fractions without uniformity in their representation. Cajori (1928, 314–335) gave one detailed historical discussion that is nevertheless incomplete. Periods and commas have been the most commonly used separators since the development of printing. Some authors, perhaps most notably John Napier of logarithm fame, even used both symbols in their work. An objection to the period is its common use to indicate multiplication, so some people have preferred centered or even raised dots, even though the same objection can be made in reverse, at least to centered dots. An objection to the comma is similarly to its common use to separate numbers in lists, although serious ambiguity should not arise so long as such lists are suitably spaced out.

Incidentally, many notations other than periods and commas have been proposed and several remained in use until well into the twentieth century. Indeed the momayyez, a mark like a forward slash or comma, is widely used at present in several Middle Eastern countries.

Stata 7 introduced `set dp comma` as a way to set the decimal point to a comma in output. Thus after

```
. set dp comma
. display 5/4
```

shows 1,25 and other output follows the same rule. Type `set dp comma, permanently` to have such output permanently. Type `set dp period` to restore the default.

Stata 7 also introduced comma-based formats such as `%7,2f`. See the help or manual entry for `format` for more details.

This still leaves the large question of input. Suppose, for example, that you have text or other data files in which numeric variables are indicated with commas as separators. Stata will not accept such variables as numeric, but the solution now is simple. Read in such variables as string, and then within Stata use `destring, replace dpcomma`. `destring` is especially convenient because it can be applied to several variables at once.

The `dpcomma` option was added to the `destring` command on 15 October 2007 and so is not documented in the Stata 10 manuals and not implemented in versions earlier than 10. Users still using Stata 9 or earlier are advised to use the `subinstr()` function to change commas to periods, followed by `destring`. Naturally, great care is required if any periods also appear as separators before or after the decimal comma. Any such periods should be removed before the comma is converted to a period.

# Reference

Cajori, F. 1928. *A History of Mathematical Notations. Volume I: Notation in Elementary Mathematics*. Chicago: Open Court.

The Stata Journal (2008)
8, Number 2, pp. 295–298

# Stata tip 62: Plotting on reversed scales

Nicholas J. Cox and Natasha L. M. Barlow
Durham University
Durham City, UK
n.j.cox@durham.ac.uk and n.l.m.barlow@durham.ac.uk

Stata has long had options allowing a reversed scale on either the $y$ or the $x$ axis of many of its graph types. Many graph users perhaps never even consider specifying such options. Those who do need to reach for them may wish to see detailed examples of how reversed scales may be exploited to good effect.

The usual Cartesian conventions are that vertical or $y$ scales increase upward, from bottom to top, and horizontal or $x$ scales increase from left to right. Vertical scales increasing downward are needed for graphs with vertical categorical axes following a table-like convention, in which the first (lowest) category is at the top of a graph, just as it would be at the top of a table. Commands such as `graph bar` and `graph dot` follow this convention. Indeed, it is likely to be so familiar that you may have to reflect briefly to see that is how such graphs are drawn. Other examples of this principle are discussed elsewhere in this issue (Cox 2008).

Reversed scales are also common in the Earth and environmental sciences. Here, in fields such as pedology, sedimentology, geomorphology, limnology, and oceanography, it is common to take measurements at varying depths within soils, sediments, rocks, and water bodies. Even though depth is not a response variable, it is conventional and convenient to plot depth below surface on the vertical axis; hence, the need for a reversed scale. An extra twist that gives spin to the graph problem is that frequently detailed analyses of materials at each level in a core, bore, or vertical profile yield several response variables, which are all to be plotted on the horizontal axis. Such multiple plotting is easiest when overlay is possible.

Let us look at some specific syntax for an example and then add comments. The data shown here come from work in progress by the second author. Sediment samples at 2 cm intervals down a core from Girdwood, Alaska, were examined for several elements associated with placer mining pollution (LaPerriere, Wagener, and Bjerklie 1985). In this example, the concentrations of gold, cadmium, arsenic, lead, copper, and zinc, measured in parts per million (ppm), are then plotted as a function of depth (cm). See figure 1.

```
. local spec clwidth(medium) msize(*0.8)

. twoway
> connect depth Au, `spec` ms(Oh) cmissing(n) ||
> connect depth Cd, `spec` ms(Th)    ||
> connect depth As, `spec` ms(Sh)    ||
> connect depth Pb, `spec` ms(O)     ||
> connect depth Cu, `spec` ms(T)     ||
> connect depth Zn, `spec` ms(S)
> yscale(reverse)  xscale(log) xscale(titlegap(*10))
> ylabel(50(10)90, angle(h)) xlabel(0.1 0.3 1 3 10 30 100)
> ytitle(Depth (cm)) xtitle(Concentration (ppm))
> legend(order(1 "Au" 2 "Cd" 3 "As" 4 "Pb" 5 "Cu" 6 "Zn") position(3) column(1))
```

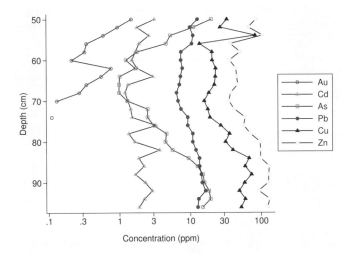

Figure 1. Variation in concentrations of various elements with depth, measured at a site in Girdwood, Alaska.

A key detail here is that Stata's model for scatter and similar plots is asymmetric. One or more $y$ variables are allowed, but only one $x$ variable, in each individual plot. Thus, if you wish to have several $x$ variables, you must either superimpose several variables, as here, or juxtapose several plots horizontally.

Further, it is necessary to spell out what is desired as text in the legend. By default, twoway would use information for the $y$ axis variables, which is not what is wanted for this kind of graph.

For these data a logarithmic scale is helpful, indeed essential, for showing several elements that vary greatly in abundance. Gold is reported as less than 0.1 ppm at depth. Such censored values cannot be shown by point symbols.

Even very experienced Stata users will not usually think up an entire graph command like this at the outset. Typically, you start with a fairly simple design and then elaborate it by a series of very small changes (which may well be changes of mind back and forth).

At some point, you are likely to find yourself transferring from the Command window to the Do-file Editor and from an interactive session to a do-file. We also find it helpful, once out of the Command window, to space out a command so that its elements are easier to see and so that it is easier to edit. Spending a few moments doing that saves some fiddly work later on. What is shown above is in fact more compressed than what typically appears within the Do-file Editor in our sessions.

The legend used here is, like all legends, at best a necessary evil, as it obliges careful readers to scan back and forth repeatedly to see what is what. The several vertical traces are fairly distinct, so one possibility is to extend the vertical axis and insert legend text at the top of the graph. '=Au[1]', for example, instructs Stata to evaluate the first value of Au and use its value. The same effect would be achieved by typing in the actual value. Here we are exploiting the shortness of the element names, which remain informative to any reader with minimal chemical knowledge. The legend itself can then be suppressed. The text elements could also be repeated at the bottom of the graph if desired. See figure 2.

```
. twoway
> connect depth Au, `spec` ms(Oh) cmissing(n) ||
> connect depth Cd, `spec` ms(Th)   ||
> connect depth As, `spec` ms(Sh)   ||
> connect depth Pb, `spec` ms(O)    ||
> connect depth Cu, `spec` ms(T)    ||
> connect depth Zn, `spec` ms(S)
> yscale(reverse r(46 .)) xscale(log) xscale(titlegap(*10))
> ylabel(50(10)90, angle(h)) xlabel(0.1 0.3 1 3 10 30 100)
> ytitle(Depth (cm)) xtitle(Concentration (ppm))
> text(48 `=Au[1]` "Au" 48 `=Cd[1]` "Cd" 48 `=As[1]` "As" 48 `=Pb[1]` "Pb"
>      48 `=Cu[1]` "Cu" 48 `=Zn[1]` "Zn")
> legend(off)
```

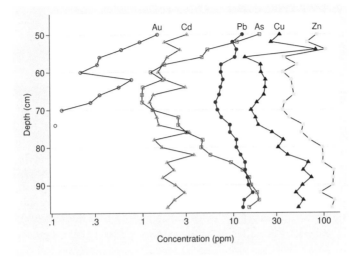

Figure 2. Variation in concentrations of various elements with depth, measured at a site in Girdwood, Alaska. The legend has been suppressed and its text elements placed within the graph.

The structure of overlays may suggest that a program be written that takes one $y$ and several $x$ variables and then works out the overlay code for you. Such a program might then be available particularly to colleagues and students less familiar with Stata. That idea is more tempting in this example than in general. In other cases, there might be a desire to mix different `twoway` types, to use different marker colors, or to make any number of other changes to distinguish the different variables. Any code flexible enough to specify any or all of that would lead to commands no easier to write than what is already possible. In computing as in cookery, sometimes you just keep old recipes once you have worked them out, at least as starting points for some later problem.

In the Earth and environmental sciences, reversed horizontal scales are also common for showing time, whenever the units are not calendar time, measured forward, but time before the present, measured backward. Within Stata, such practice typically requires no more than specifying `xscale(reverse)`.

# References

Cox, N. J. 2008. Speaking Stata: Between tables and graphs. *Stata Journal* 8: 269–289.

LaPerriere, J. D., S. M. Wagener, and D. M. Bjerklie. 1985. Gold-mining effects on heavy metals in streams, Circle Quadrangle, Alaska. *Journal of the American Water Resources Association* 21: 245–252.

The Stata Journal (2008)
8, Number 2, pp. 299–303

# Stata tip 63: Modeling proportions

Christopher F. Baum
Department of Economics
Boston College
Chestnut Hill, MA
baum@bc.edu

You may often want to model a response variable that appears as a proportion or fraction: the share of consumers' spending on food, the fraction of the vote for a candidate, or the fraction of days when air pollution is above acceptable levels in a city. To handle these data properly, you must take account of the bounded nature of the response. Just as a linear probability model on unit record data can generate predictions outside the unit interval, using a proportion in a linear regression model will generally yield nonsensical predictions for extreme values of the regressors.

One way to handle this for response variables' values strictly within the unit interval is the logit transformation

$$y = \frac{1}{1 + \exp(-X\beta)}$$

which yields the transformed response variable $y^*$

$$y^* = \log\left(\frac{y}{1-y}\right) = X\beta + \epsilon$$

where we have added a stochastic error process $\epsilon$ to the model to be fitted. This transformation may be performed with Stata's logit() function. We can then use linear regression ([R] **regress**) to model $y^*$, the *logit transformation* of $y$, as a linear function of a set of regressors, $X$. If we then generate predictions for our model ([R] **predict**), we can apply Stata's invlogit() function to express the predictions in units of $y$. For instance,

```
. use http://www.stata-press.com/data/r10/census7
(1980 Census data by state)
. generate adultpop = pop18p/pop
. quietly tabulate region, generate(R)
. generate marrate = marriage/pop
. generate divrate = divorce/pop
. generate ladultpop = logit(adultpop)
```

```
. regress ladultpop marrate divrate R1-R3
```

| Source | SS | df | MS | | Number of obs = | 49 |
|---|---|---|---|---|---|---|
| | | | | | F( 5, 43) = | 4.33 |
| Model | .164672377 | 5 | .032934475 | | Prob > F = | 0.0028 |
| Residual | .327373732 | 43 | .007613343 | | R-squared = | 0.3347 |
| | | | | | Adj R-squared = | 0.2573 |
| Total | .492046109 | 48 | .010250961 | | Root MSE = | .08725 |

| ladultpop | Coef. | Std. Err. | t | P>\|t\| | [95% Conf. Interval] | |
|---|---|---|---|---|---|---|
| marrate | -18.26494 | 7.754941 | -2.36 | 0.023 | -33.90427 | -2.625615 |
| divrate | 8.600844 | 12.09833 | 0.71 | 0.481 | -15.79777 | 32.99946 |
| R1 | .1192464 | .0482428 | 2.47 | 0.017 | .0219555 | .2165373 |
| R2 | .0498657 | .042209 | 1.18 | 0.244 | -.0352569 | .1349883 |
| R3 | .0582061 | .0357729 | 1.63 | 0.111 | -.0139368 | .130349 |
| _cons | .999169 | .093568 | 10.68 | 0.000 | .8104712 | 1.187867 |

```
. predict double ladultpophat, xb

. generate adultpophat = invlogit(ladultpophat)

. summarize adultpop adultpophat
```

| Variable | Obs | Mean | Std. Dev. | Min | Max |
|---|---|---|---|---|---|
| adultpop | 49 | .7113268 | .0211434 | .6303276 | .7578948 |
| adultpophat | 49 | .7116068 | .0120068 | .6855482 | .7335103 |

Alternatively, we could use Stata's grouped logistic regression ([R] **glogit**) to fit the model. This command uses the same transformation on the response variable, which must be provided for the number of positive responses and the total number of responses (that is, the numerator and denominator of the proportion). For example,

```
. glogit pop18p pop marrate divrate R1-R3
Weighted LS logistic regression for grouped data
```

| Source | SS | df | MS | | Number of obs = | 49 |
|---|---|---|---|---|---|---|
| | | | | | F( 5, 43) = | 4.24 |
| Model | .129077492 | 5 | .025815498 | | Prob > F = | 0.0032 |
| Residual | .261736024 | 43 | .006086884 | | R-squared = | 0.3303 |
| | | | | | Adj R-squared = | 0.2524 |
| Total | .390813516 | 48 | .008141948 | | Root MSE = | .07802 |

| pop18p | Coef. | Std. Err. | t | P>\|t\| | [95% Conf. Interval] | |
|---|---|---|---|---|---|---|
| marrate | -22.82454 | 8.476256 | -2.69 | 0.010 | -39.91853 | -5.730537 |
| divrate | 18.44877 | 12.66291 | 1.46 | 0.152 | -7.088418 | 43.98596 |
| R1 | .0762246 | .0458899 | 1.66 | 0.104 | -.0163212 | .1687704 |
| R2 | -.0207864 | .0362001 | -0.57 | 0.569 | -.0937909 | .0522181 |
| R3 | .0088961 | .0354021 | 0.25 | 0.803 | -.062499 | .0802912 |
| _cons | 1.058316 | .0893998 | 11.84 | 0.000 | .8780241 | 1.238608 |

These results differ from those of standard regression because `glogit` uses weighted least-squares techniques. As explained in [R] **glogit**, the appropriate weights correct for the heteroskedastic nature of $\epsilon$ has zero mean but variance equal to

$$\sigma_j^2 = \frac{1}{n_j p_j (1 - p_j)}$$

By generating those weights, where $n_j$ is the number of responses in the $j$th category and $p_j$ is the predicted value we computed above, we can reproduce the `glogit` results with `regress` by using analytic weights, as verified with the commands:

```
. generate glswt = adultpophat * (1 - adultpophat) * pop
. quietly regress ladultpop marrate divrate R1-R3 [aw=glswt]
```

In the case of these state-level census data, values for the proportion $y$ must lie within the unit interval. But we often consider data for which the limiting values of zero or one are possible. A city may spend 0% of its budget on preschool enrichment programs. A county might have zero miles of active railway within its boundaries. There might have been zero murders in a particular town in each of the last five years. A hospital may have performed zero heart transplants last year. In other cases, we may find values of one for particular proportions of interest. Neither zeros nor ones can be included in the strategy above, as the logit transformation is not defined for those values.

A strategy for handling proportions data in which zeros and ones may appear as well as intermediate values was proposed by Papke and Wooldridge (1996). At the time of their writing, Stata's generalized linear model ([R] **glm**) command could not handle this model, but it has been enhanced to do so. This approach makes use of the logit link function (that is, the logit transformation of the response variable) and the binomial distribution, which may be a good choice of family even if the response is continuous. The variance of the binomial distribution must go to zero as the mean goes to either 0 or 1, as in each case the variable is approaching a constant, and the variance will be maximized for a variable with mean of 0.5.

To illustrate, consider an alternative dataset that contains zeros and ones in its response variable, `meals`: the proportion of students receiving free or subsidized meals at school.

```
. use http://www.ats.ucla.edu/stat/stata/faq/proportion, clear

. summarize meals

    Variable |        Obs        Mean    Std. Dev.        Min        Max
-------------+--------------------------------------------------------
       meals |       4421    .5188102    .3107313          0          1

. glm meals yr_rnd parented api99, link(logit) family(binomial) vce(robust) nolog
note: meals has noninteger values

Generalized linear models                          No. of obs      =       4257
Optimization     : ML                              Residual df     =       4253
                                                   Scale parameter =          1
Deviance         =  395.8141242                    (1/df) Deviance =    .093067
Pearson          =  374.7025759                    (1/df) Pearson  =   .0881031

Variance function: V(u) = u*(1-u/1)                [Binomial]
Link function    : g(u) = ln(u/(1-u))              [Logit]
                                                   AIC             =   .7220973
Log pseudolikelihood = -1532.984106                BIC             =  -35143.61

-------------------------------------------------------------------------------
             |               Robust
       meals |      Coef.   Std. Err.      z    P>|z|     [95% Conf. Interval]
-------------+-----------------------------------------------------------------
      yr_rnd |   .0482527   .0321714     1.50   0.134    -.0148021    .1113074
    parented |  -.7662598   .0390715   -19.61   0.000    -.8428386   -.6896811
       api99 |  -.0073046   .0002156   -33.89   0.000    -.0077271   -.0068821
       _cons |    6.75343   .0896767    75.31   0.000     6.577667    6.929193
-------------------------------------------------------------------------------
```

The techniques used above can be used to generate predictions from the model and transform them back into the units of the response variable. This approach is preferred to that of dropping the observations with zero or unit values, which would create a truncation problem, or coding them with some arbitrary value ("winsorizing") such as 0.0001 or 0.9999.

Some researchers have considered using censored normal regression techniques such as tobit ([R] **tobit**) on proportions data that contain zeros or ones. However, this is not an appropriate strategy, as the observed data in this case are not censored: values outside the [0, 1] interval are not feasible for proportions data.

One concern was voiced about proportions data containing zeros or ones.[1] In the context of the generalized tobit or "heckit" model ([R] **heckman**), we allow for limit observations (for instance, zero values) being generated by a different process than non-censored observations. The same argument may apply here, in the case of proportions data: the managers of a city that spends none of its resources on preschool enrichment programs have made a discrete choice. A hospital with zero heart transplants may be a facility whose managers have chosen not to offer certain advanced services.

In this context, the glm approach, while properly handling both zeros and ones, does not allow for an alternative model of behavior generating the limit values. If different factors generate the observations at the limit points, a sample selection issue arises. Li and Nagpurnanand (2007) argue that selection issues arise in numerous variables of interest in corporate finance research. In a forthcoming article, Cook, Kieschnick, and

---

1. See, for instance, McDowell and Cox (2001).

McCullough (2008) address this issue for proportions of financial variables by developing what they term the "zero-inflated beta" model, which allows for zero values (but not unit values) in the proportion and for separate variables influencing the zero and nonzero values.[2]

# References

Cook, D. O., R. Kieschnick, and B. D. McCullough. 2008. Regression analysis of proportions in finance with self selection. *Journal of Empirical Finance* 15: 860–867.

Li, K., and R. Nagpurnanand. 2007. Self-selection models in corporate finance. In *Handbook of Corporate Finance: Empirical Corporate Finance*, ed. B. E. Eckbo, chap. 2. Amsterdam: Elsevier.

McDowell, A., and N. J. Cox. 2001. FAQ: How do you fit a model when the dependent variable is a proportion? http://www.stata.com/support/faqs/stat/logit.html.

Papke, L. E., and J. M. Wooldridge. 1996. Econometric methods for fractional response variables with an application to 401(K) plan participation rates. *Journal of Applied Econometrics* 11: 619–632.

---

2. Their approach generalizes models fit with the beta distribution; user-written programs for that purpose may be located by typing `findit beta distribution`.

The Stata Journal (2008)
**8**, Number 3, pp. 444–445

# Stata tip 64: Cleaning up user-entered string variables[1]

Jeph Herrin
Yale School of Medicine
Yale University
New Haven, CT
jeph.herrin@yale.edu

Eva Poen
School of Economics
University of Nottingham
Nottingham, UK
eva.poen@gmail.com

A common problem in data management, especially when using large databases that receive entries from many different people, is that the same name is given in several different forms. This problem can arise in many instances, for example, lists of names of schools, hospitals, drugs, companies, countries, and so forth. Variation can reflect several genuinely different forms of the same name as well as a multitude of small errors or idiosyncrasies in spelling, punctuation, spacing, and use of uppercase or lowercase.

Thus, in context, a person may have no difficulty in recognizing that values of a string variable, such as "New York", "New York City", "N Y C", and so on, all mean the same thing. However, a program like Stata is necessarily literal and will treat them as distinct. How do we massage data so that values with the same meaning are represented in the same way? Several techniques exist for these purposes. Here we outline a simple strategy for ensuring that names are as tidy as possible.

As a preliminary stage, it is useful to try to eliminate small inconsistencies before you look at the individual observations. A good tactic is to keep the original names in one variable, exactly as given, and to work with one or more variables that contain cleaned-up versions. Some common problems are the following:

- Leading and trailing spaces may not be evident but will cause Stata to treat values as distinct. Thus "New York City" and "New York City " are not considered equal by Stata until `trim()` is used to delete the trailing space.

- Similarly, inconsistencies in internal spacing can cause differences that Stata will register. The `itrim()` function will reduce multiple, consecutive internal blanks to single internal blanks.

- Variations of uppercase and lowercase can also be troublesome. The `upper()`, `lower()`, or `proper()` functions can be used to make names consistent.

- Other common differences include whether hyphens are present, whether accented characters appear with or without accents or in some other form, and whether ampersands are printed as characters or as equivalent words.

- A large class of problems concerns abbreviations.

In the last two cases, `subinstr()` is a useful function for making changes toward consistent conventions. Note a common element here: string functions, documented in [D] **functions**, are invaluable for cleaning up strings.

---

1. This tip was updated to reflect the new `merge` syntax.—Ed.

After a preliminary cleaning, you can create a list of all the names that you have. Usually, this list is shorter than the number of observations. It is worthwhile to inspect the list of names and look for further clean-up possibilities before proceeding. A tabulation of names, say, by using `tabulate, sort`, serves this purpose and provides you with the number of occurrences for each variation.

After the cleaning is completed, you are ready to compile your list of names. Suppose that the `name` variable contains the names.

```
. use mydatafile
. by name, sort: keep if _n == 1
. keep name
```

Now open the Data Editor by typing

```
. edit name
```

and add a second—numeric—variable, say, `code`. In this second variable, give the same number for every observation that represents the same object. This will be moderately time consuming, but because data are sorted on `name`, it may just take a few minutes.

Now exit the Data Editor, sort on `name`, and save the dataset:

```
. sort name
. save codes, replace
```

Some people may prefer to create the code in their favorite spreadsheet or text editor, say, if an outside expert not adept at Stata is recruited to do the coding. The principles are the same: you need to export the data to the other application and then read data back into Stata. You may lose out on an audit trail if the other software does not offer an equivalent to a Stata `.log` file.

Now you have a file (`codes.dta`) that has a list of the names in all their variety, as well as a set of numeric codes, which you can return to later to check your work. The key thing is that you can now `merge` this file into your original file to assign a common code to every value of `name` that is the same:

```
. use mydatafile, clear
. sort name
. merge m:1 name using codes
```

As always when using `merge`, examine the `_merge` variable; here, if `_merge` is not always equal to 3, then you have made a mistake somewhere. You should also examine `code`; if there are any missing values, you will need to `edit` the file `codes.dta` again to add them.

Now you can identify the objects by their codes; if you want, you can assign a common name:

```
. by code, sort: replace name = name[1]
```

The Stata Journal (2008)
**8**, Number 3, pp. 446–447

# Stata tip 65: Beware the backstabbing backslash

Nicholas J. Cox
Department of Geography
Durham University
Durham City, UK
n.j.cox@durham.ac.uk

The backslash character, \, has two main roles for Stata users who have Microsoft Windows as their operating system. This tip warns you to keep these roles distinct, because Stata's interpretation of what you want may puzzle you. The problem is signaled at [U] **18.3.11 Constructing Windows filenames by using macros** but nevertheless bites often enough that another warning may be helpful.

The first and better known role is that the backslash acts as a separator in full specifications of directory or filenames. Thus, on many Windows machines, a Stata executable may be found within C:\Program Files\Stata10. Note the two backslashes in this example.

The second and lesser known role is that the backslash is a so-called escape character that suppresses the default interpretation of a character.

The best example in Stata is that the left quotation mark, `, is used to delimit the start of local macro names. Thus `frog` to Stata is a reference to a local macro called frog. Typically, Stata sees such a reference and substitutes the contents of the local macro frog at the same place. If no such macro is visible, that is not a bug. Instead, Stata substitutes an empty string for the macro name.

What happens if you want the conventional interpretation of the left quotation mark? You use a backslash to flag that you want to override the usual Stata interpretation.

```
. display "He said \`frog´."
He said `frog´.
```

It is perhaps unlikely, although clearly not impossible, that you do want this in practice. The problem is that you may appear to Stata to specify this, even though you are likely to do it purely by accident.

Suppose, for example, that you are looping over a series of datasets, reading each one into Stata, doing some work, and then moving on to the next.

Your code may look something like this:

```
. foreach f in a b c {
.     use "c:\data\this project\`f´"
.     and so on
. }
```

Do you see the difficulty now? Your intent is that the loop uses a local macro, which in turn takes the values a, b, and c to read in the datasets a.dta, b.dta, and c.dta. But Stata sees the last backslash as an instruction to escape the usual interpretation of the left quotation mark character that immediately follows. The resulting misunderstanding will crash your code.

Any Unix (including Macintosh) users reading this will feel smug, because they use the forward slash, /, within directory or filenames, and this problem never bites them. The way around the problem is by knowing that Stata will let you do that too, even under Windows. Stata takes it upon itself to translate between you and the operating system. You certainly need to use the forward slash in this example to escape the difficulty of the implied escape. Typing `use "c:\data\this project/'f'"` would solve the problem. A backslash is only problematic whenever it can be interpreted as an escape character. In fact, you can mix forward and backward slashes willy-nilly in directory or filenames within Stata for Windows, so long as you do not use a backslash just before a left quotation mark.

The tidiest solution is to use forward slashes consistently, as in `use "c:/data/this project/'f'"`, although admittedly this may clash strongly with your long-practiced Windows habits.

The Stata Journal (2008)
8, Number 3, pp. 448–449

# Stata tip 66: ds—A hidden gem[1]

Martin Weiss
University of Tuebingen
Tuebingen, Germany
martin.weiss@uni-tuebingen.de

ds is one of a few dozen "undocumented" commands in Stata whose names are available in `help undocumented`. Contrary to the help file assertion that "an undocumented command is a command of very limited interest, usually only to Stata programmers", ds is extremely helpful, both interactively and in programs. The main hurdle to its widespread adoption by Stata users seems to be limited awareness of its existence.

Stata users are generally familiar with the `describe` (see [D] **describe**) command. `describe` allows you to gain a rapid overview of the dataset in memory, or with the `using` modifier, a dataset residing on a hard disk or available on the Internet. ds also allows that with its `detail` option, omitting only the general information provided in the header of the output of the `describe` command. For example,

```
. sysuse uslifeexp2.dta
(U.S. life expectancy, 1900-1940)

. describe

Contains data from C:\Program Files\Stata10\ado\base\u\uslifeexp2.dta
  obs:            41                          U.S. life expectancy, 1900-1940
  vars:            2                          2 Apr 2007 14:39
  size:          574 (99.9% of memory free)   (_dta has notes)

              storage   display    value
variable name   type    format     label      variable label

year            int     %9.0g                 Year
le              float   %9.0g                 life expectancy

Sorted by:  year

. ds, detail

              storage   display    value
variable name   type    format     label      variable label

year            int     %9.0g                 Year
le              float   %9.0g                 life expectancy
```

More importantly, ds also provides the means of identifying subsets of variables with specified properties. It complements `lookfor` (see [D] **lookfor**), which allows you to search for certain strings in variable names or variable labels. ds enhances this functionality extensively, letting you specify

---

1. As of Stata 11, ds is a "documented" command; see [D] **ds**.—Ed.

- certain types of variables;

- whether variable labels, value labels, and characteristics have been attached, and if so, whether they match certain patterns; and

- variables with specific formats.

A further systematic feature is that you can specify either the subset of variables satisfying particular properties or the complementary subset that does not satisfy those properties. As a simple example of the latter, when you are using the `auto.dta`, the command `ds make, not` specifies all variables other than `make`.

These capabilities prove particularly handy with large or poorly known datasets. As a simple example, pretend you were not familiar with the `auto` dataset and were looking for string variables.

```
. sysuse auto.dta
(1978 Automobile Data)
. * show all variables featuring type string
. ds, has(type string)
make
```

While `describe` would list the variable types, leaving the task of finding a certain type to you, `ds` can provide precisely what you were looking for. Despite its "undocumented" status, a dialog box can ease navigation through the intricacies of this command: to try the dialog box, type `db ds`. Beyond the results shown as usual, `ds` also leaves behind a list of variables found in `r(varlist)`, which is available for use by subsequent commands, such as `list` (see [D] **list**) or `summarize` (see [R] **summarize**). Many of the properties that you can search for with `ds` can also be extracted with extended macro functions; see [P] **macro**. To illustrate, consider the `voter` dataset shipped with Stata.

```
. sysuse voter.dta
. * show all variables with value labels attached
. ds, has(vall)
candidat  inc
. * show all variables not of type float
. ds, not(type float)
candidat  inc       pfrac     pop
. * show mean of all variables with format %10.0g
. ds, has(format %10.0g)
pfrac  pop
. tabstat `r(varlist)'
```

| stats | pfrac | pop |
|-------|-------|-----|
| mean | 6.733333 | 104299.3 |

The help file accessed by typing `help ds` gives several more examples. The help file accessed by typing `help varfind_pat_examp` explains the use of wildcards within the specification of patterns.

The Stata Journal (2008)
**8**, Number 3, pp. 450–451

# Stata tip 67: J() now has greater replicating powers

Nicholas J. Cox
Department of Geography
Durham University
Durham City, UK
n.j.cox@durham.ac.uk

The Mata standard function J() was generalized in the Stata update of 25 February 2008. This tip flags its greater replicating powers. Note that J() in Stata's original matrix language remains as it was.

Users of Mata will have become accustomed to the role of J() in creating matrices of constants. For example, once within Mata,

```
: J(5,5,1)
[symmetric]
        1    2    3    4    5
    1   1
    2   1    1
    3   1    1    1
    4   1    1    1    1
    5   1    1    1    1    1
```

You may be used to thinking of the way J() works like this: I want a $5 \times 5$ matrix, all of whose elements are the scalar 1. Another way of thinking about it is this: Give me 5 replicates or copies rowwise and 5 copies columnwise of the scalar 1. The results are identical when scalars are being replicated.

The second way of thinking about it helps in understanding the generalization now in place. What is to be replicated can now be a matrix, naturally including not only scalars but also vectors as special cases.

The help file gives full technical details and a variety of examples, but here is another. My Speaking Stata column in this issue (Cox 2008) mentions the bias on Fisher's $z$ scale when estimating correlation $r$ from sample size $n$ of $2r/(n-1)$. The question is thus how big this is for a variety of values of $r$ and $n$. We can quickly get a table from Mata:

```
: r = J(5, 1, (.1, .3, .5, .7, .9))

: r
        1    2    3    4    5
    1   .1   .3   .5   .7   .9
    2   .1   .3   .5   .7   .9
    3   .1   .3   .5   .7   .9
    4   .1   .3   .5   .7   .9
    5   .1   .3   .5   .7   .9
```

```
: n = J(1, 5, (10, 20, 50, 100, 200)´)

: n
          1      2      3      4      5
   ┌─────────────────────────────────────┐
 1 │     10     10     10     10     10   │
 2 │     20     20     20     20     20   │
 3 │     50     50     50     50     50   │
 4 │    100    100    100    100    100   │
 5 │    200    200    200    200    200   │
   └─────────────────────────────────────┘
```

```
: 2 * r :/ (n :- 1)
                 1               2               3               4               5
   ┌──────────────────────────────────────────────────────────────────────────────────┐
 1 │   .0222222222    .0666666667    .1111111111    .1555555556              .2          │
 2 │   .0105263158    .0315789474    .0526315789    .0736842105     .0947368421          │
 3 │   .0040816327     .012244898    .0204081633    .0285714286     .0367346939          │
 4 │    .002020202    .0060606061    .0101010101    .0141414141     .0181818182          │
 5 │   .0010050251    .0030150754    .0050251256    .0070351759     .0090452261          │
   └──────────────────────────────────────────────────────────────────────────────────┘
```

Notice again how the first two arguments of J() are the numbers of replicates or copies, rowwise and columnwise, and not necessarily the numbers of rows and columns in the resulting matrix.

## Reference

Cox, N. J. 2008. Speaking Stata: Correlation with confidence, or Fisher's z revisited. *Stata Journal* 8: 413–439.

The Stata Journal (2010)
**10**, Number 4, pp. 682–685

# Stata tip 68: Week assumptions

Nicholas J. Cox
Department of Geography
Durham University
Durham, UK
n.j.cox@durham.ac.uk

## 1   Introduction

Stata's handling of dates and times is centered on daily dates. Days are aggregated into weeks, months, quarters, half-years, and years. Days are divided into hours, minutes, and seconds. This may all sound simple in principle, leaving just the matter of identifying the syntax for converting from one form of representation to another. For an introduction, see [U] **24 Working with dates and times**. For more comprehensive treatments, see [D] **datetime** and [D] **functions**. As a matter of history, know that specific date functions were introduced in Stata 4 in 1995, replacing an earlier system based on ado-files. These date functions were much enhanced in Stata 6 in 1999 and again in Stata 10 in 2007.

However, matters are not quite as simple as this description implies. The occasional addition of leap seconds lengthening the year is a complication arising with some datasets. A little thought shows that weeks are also awkward—whatever the precise definition of a week, weeks are not guaranteed to nest neatly and evenly into months, quarters, half-years, or years. This tip focuses on Stata's solution for weeks and on how to set up your own alternatives given different definitions of the week. A conventional Western calendar with seven-day weeks is assumed throughout this tip. For much richer historical, cultural, and computational context, see Richards (1998), Holford-Strevens (2005), and Dershowitz and Reingold (2008).

Gabriel Rossman is thanked for providing stimulating comments.

## 2   Stata's definition of weeks

Stata's definition of weeks matches one desideratum and violates others, as would any other definition. For Stata, week 1 of any year starts on 1 January, whatever day of the week that is (Sunday through Saturday).

```
. display %tw wofd(mdy(1,1,2010))
  2010w1
. display %tw wofd(mdy(1,7,2010))
  2010w1
. display %tw wofd(mdy(1,8,2010))
  2010w2
```

In these examples, two date functions are used: `mdy()` yields daily dates from month, day, and year components, and `wofd()` converts such dates to the corresponding weeks. Most users prefer to see dates shown intelligibly with date display formats such as `%tw`, thus hiding the underlying Stata machinery, which pivots on a convention that 1 January 1960 is date origin or day 0.

At the end of the year, for Stata the 52nd week always lasts 8 days in nonleap years and 9 days in leap years. (Recall that $52 \times 7 = 364$, so a calendar year always has either 1 day or 2 days more than that.)

This definition ensures that weeks nest within years, meaning that no week ever starts in one calendar year and finishes in the next. Also by this definition, a year has precisely 52 weeks, even though the last week is never 7 days long. Naturally, this solution (and indeed any other) cannot ensure that weeks always nest within months, quarters, or half-years. (February is sometimes an exception, but necessarily the only one.)

## 3    Alternative assumptions

Users who deal with weeks may wish to work with other definitions. The most obvious alternatives arise from regarding particular days of the week as defining either the start or the end of the week. With these definitions, the relation of weeks to the days of the week is fixed, and all weeks last precisely 7 days. However, other desiderata are violated: notably, weeks may now span two calendar years.

Such definitions are most likely to appeal whenever weekly cycles are part of what is being investigated and particular days have meaning. Examples from major religions will be familiar. Particular days of the week are often key for financial transactions. More parochially, Durham University teaching weeks start on Thursdays in the first term of each academic year and on Mondays in the other two terms.

With such alternatives, there is no need to set up a new numbering system for weeks, at least as far as working within Stata is concerned. Each week can just be identified by its start or end date, whichever is desired. If you want to map the weeks that do occur in your dataset to a simple numbering scheme, `egen, group()` does this easily.

Classifying weeks by their start days is an exercise in rounding down, and classifying them by their end days is one in rounding up, so solutions using `floor()` and `ceil()` are possible; see Cox (2003). However, a solution using a date is likely to seem more attractive.

Consider this pretend dataset of 8 days in 2010.

```
. clear
. set obs 8
obs was 0, now 8
. gen day = _n + mdy(9,25,2010)
. gen day2 = day
```

```
. format day %tdDay
. format day2 %tdd_m
. gen dow = dow(day)
. list
```

|      | day  | day2   | dow |
|------|------|--------|-----|
| 1.   | Sun  | 26 Sep | 0   |
| 2.   | Mon  | 27 Sep | 1   |
| 3.   | Tue  | 28 Sep | 2   |
| 4.   | Wed  | 29 Sep | 3   |
| 5.   | Thu  | 30 Sep | 4   |
| 6.   | Fri  | 1 Oct  | 5   |
| 7.   | Sat  | 2 Oct  | 6   |
| 8.   | Sun  | 3 Oct  | 0   |

The Stata function `dow()` returns day of the week coded 0 for Sunday through 6 for Saturday. The dates 26 September 2010 and 3 October 2010 were Sundays. Now we have within reach one-line solutions for various problems, given here in terms of the example daily date variable `day`.

The simplest case is classifying days in each week by the Sundays that start them. Glancing at the listing above shows that we just need to subtract `dow(day)` from `day`:

```
. gen sunstart = day - dow(day)
```

Working with other days can be thought of as rotating the days of the weeks to an origin other than Sunday. One good way to do that is using the versatile function `mod()` (Cox 2007). If $d$ is one of $1, \ldots, 6$, then `mod(dow(day) - d, 7)` rotates the results of `dow()` so that day $d$ is the new origin. $d = 0$ leaves the days as they were.

Hence if we wish to classify days by starting Fridays, subtract `mod(dow(day) - 5, 7)` from `day`:

```
. gen fristart = day - mod(dow(day) - 5, 7)
```

This is consistent with the simpler rule for Sunday. As just implied, `mod(dow(day) - 0, 7)` is identical to `mod(dow(day), 7)` and in turn to `dow(day)`.

Now consider classifying weeks by the days that end them. We now need to add an appropriate number between 0 and 6 to the date. That number will cycle from 6 to 0 to 6, rather than from 0 to 6 to 0; it is given by `mod(d - dow(day), 7)`. So, Friday-ending weeks are given by

```
. gen friend = day + mod(5 - dow(day), 7)
```

This trick also offers a good way to generate ending Sundays:

```
. gen sunend = day + mod(0 - dow(day), 7)
```

The zero in the line just above is not necessary but is left in to emphasize the resemblance to the previous line.

Finally, let's show the results of these calculations. There are several other ways to compute them, but using `mod()` has its attractions. This method could also be extended to weeks that are not 7 days long, which do arise in certain research problems.

```
. format su* fr* %tdd_m
. list day* *start *end
```

|   | day | day2 | sunstart | fristart | friend | sunend |
|---|-----|------|----------|----------|--------|--------|
| 1. | Sun | 26 Sep | 26 Sep | 24 Sep | 1 Oct | 26 Sep |
| 2. | Mon | 27 Sep | 26 Sep | 24 Sep | 1 Oct | 3 Oct |
| 3. | Tue | 28 Sep | 26 Sep | 24 Sep | 1 Oct | 3 Oct |
| 4. | Wed | 29 Sep | 26 Sep | 24 Sep | 1 Oct | 3 Oct |
| 5. | Thu | 30 Sep | 26 Sep | 24 Sep | 1 Oct | 3 Oct |
| 6. | Fri | 1 Oct | 26 Sep | 1 Oct | 1 Oct | 3 Oct |
| 7. | Sat | 2 Oct | 26 Sep | 1 Oct | 8 Oct | 3 Oct |
| 8. | Sun | 3 Oct | 3 Oct | 1 Oct | 8 Oct | 3 Oct |

# References

Cox, N. J. 2003. Stata tip 2: Building with floors and ceilings. *Stata Journal* 3: 446–447.

———. 2007. Stata tip 43: Remainders, selections, sequences, extractions: Uses of the modulus. *Stata Journal* 7: 143–145.

Dershowitz, N., and E. M. Reingold. 2008. *Calendrical Calculations*. 3rd ed. Cambridge: Cambridge University Press.

Holford-Strevens, L. 2005. *The History of Time: A Very Short Introduction*. Oxford: Oxford University Press.

Richards, E. G. 1998. *Mapping Time: The Calendar and Its History*. Oxford: Oxford University Press.

**Editors' note.** Some alert readers have noted that there was no Stata tip 68; we moved from 67 to 69 without explanation. This was pure oversight on the part of the *Stata Journal*. In Stata tradition, we now fix this bug with an apology, particularly so that any future compilations of tips include precisely the number advertised.

The Stata Journal (2008)
**8**, Number 4, pp. 583–585

# Stata tip 69: Producing log files based on successful interactive commands

Alan R. Riley
StataCorp
College Station, TX
ariley@stata.com

So, your interactive Stata session went well and you got some good results. Naturally, you made sure you kept a log file by using the `log` command ([R] **log**).

But, almost inevitably, you also made some errors in your commands. And perhaps you also have within your log some digressions, repetitions, or things that turned out to be not so interesting or useful. How do you now produce a log file based only on the successful commands? More importantly, how do you save the sequence of commands you issued so that you can reproduce your results?

Such questions are longstanding, and there are possible software solutions on various levels. You might reach for the Stata Do-file Editor or your favorite text editor or scripting language, or you might write a program using Stata commands such as `file` ([P] **file**) to edit the log file down to the valuable part. See, for example, Cox (1994) or Eng (2007) for some detailed suggestions.

Here I concentrate on two approaches that should help.

The first approach makes a good start by using Stata's Review window, which displays a history of commands submitted to Stata. The following steps will save a do-file consisting of all the interactive commands that did not result in an error.

There are three columns in the Review window: sequence number (the order in which commands were submitted to Stata), command (the command itself), and return code (the return code, or `_rc`, from the command if it exited with an error; this column is empty if the command completed successfully).

After issuing several commands to Stata interactively, some of which might have resulted in errors, click on the top of the return code (`_rc`) column in the Review window. This will sort the commands in the Review window: now all the commands that did not result in an error are grouped together (and within that group they will be in the order in which they were submitted to Stata). Beneath them will be all the commands resulting in errors, sorted by the return code. The screenshots below show what the Review window might look like before and after doing this.

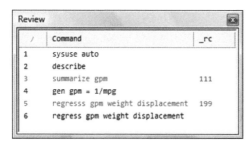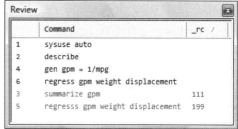

The first group is of interest; the second can be ignored. Select all the commands in the first group: click once on the first command in the group to select it, scroll until the last command in the group is visible, and then hold down the *Shift* key while clicking on that last command to select all the commands in the group.

Once all the valid commands have been selected, right-click anywhere in the Review window and choose **Save Selected...** to save those commands to a do-file, or choose **Send to Do-file Editor** to paste those commands into Stata's Do-file Editor.

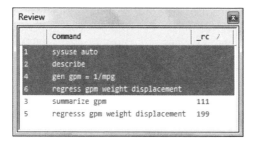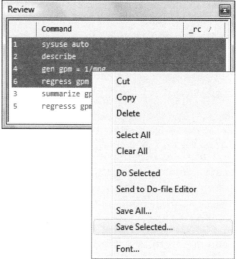

Naturally, this does not solve the questions of digressions, repetitions, or less worthwhile results. Nor is there an absolute guarantee that rerunning these commands would result in exactly the same state as you ended your session. There is a chance that some of the commands that produced an error changed your data, for example, if one of your commands was to run a do-file that stopped with an error after making certain changes to your data. However, knowing this approach should make your task easier.

The second approach uses `cmdlog`. Make sure that you run both `log` and `cmdlog` simultaneously. Then, with an eye on the log, edit the command log so that it contains only commands that were legal and useful. Then run the command log as a do-file to get the error-free log. In essence, there is less work to do that way.

# References

Cox, N. J. 1994. os13: Using awk and fgrep for selective extraction from Stata log files. *Stata Technical Bulletin* 19: 15–17. Reprinted in *Stata Technical Bulletin Reprints*, vol. 4, pp. 78–80. College Station, TX: Stata Press.

Eng, J. 2007. File filtering in Stata: Handling complex data formats and navigating log files efficiently. *Stata Journal* 7: 98–105.

The Stata Journal (2008)
8, Number 4, pp. 586–587

# Stata tip 70: Beware the evaluating equal sign[1]

Nicholas J. Cox
Department of Geography
Durham University
Durham, UK
n.j.cox@durham.ac.uk

When you assign to a local or global macro the result of evaluating an expression, you can lose content because of limits on the length of string expression that Stata will handle. This tip explains this pitfall and the way around it in detail. The problem is described in [U] **18.3.4 Macros and expressions**, but it occurs often enough that another warning may be helpful. Illustrations will be in terms of local macros, but the warning applies equally to global macros.

What is the difference between the following two statements?

```
. local foo1 "just 23 characters long"
. local foo2 = "just 23 characters long"
```

Let's look at the results:

```
. display "`foo1'"
just 23 characters long
. display "`foo2'"
just 23 characters long
```

The answer is nothing, in terms of results. Do not let that deceive you. Two quite different processes are involved.

The first statement defines local macro `foo1` by copying the string `just 23 characters long` into it. The second statement does more. It first evaluates the string expression to the right of the equal sign and then defines the local macro by assigning the result of that evaluation to it.

The mantra to repeat is "the equal sign implies evaluation". Here the evaluation makes no difference to the result, but that will be not true in other cases. A more treacherous pitfall lies ahead.

It should be clear that we often need an evaluation. If the statements had been

```
. local bar1 lower("FROG")
. local bar2 = lower("FROG")
```

the results would have been quite different:

```
. display `""`bar1'""'
lower("FROG")
```

---

1. As of Stata 13, the limit on the length of a string in a string expression in all flavors of Stata has increased to 2 billion.—Ed.

```
. display "`bar2'"
frog
```

The first just copies what is on the right to what is named on the left, the local macro `bar1`. The second does the evaluation, and then the result is put in the local macro.

Whenever you need an evaluation, you should be aware of a potential limit. `help limits` gives you information on limits applying to your Stata. As seen in the list, there are many different limits, and they can vary between Stata version and Stata flavor; there is no point in echoing that list here. But at the time of writing, the limit on the length of a string in a string expression in all flavors is 244 characters, which is quite a modest number.

The implication is that whenever you have a choice between copying and evaluation, always use copying. And if you must use evaluation, make sure that no single evaluation is affected by the limit of 244 characters. Further, be aware of work-arounds. Thus the `length()` function cannot report string lengths above 245 (yes, 245) characters, but the extended function `: length local` can measure much longer strings. See `help extended_fcn` for more details.

If you do not pay attention to the limit, you may run into problems. First, your local macro will be shorter than you want. Second, it may not even make sense to whatever you feed it to. Any error message you see will then not refer to the real underlying error, a truncated string. So your bug may be elusive.

For example, users sometimes collect variable names (or other names) in a list by using a loop centered on something like this:

```
. local mylist = "`mylist' `newitem'"
```

This may seem like self-evidently clear and correct code. But as `mylist` grows beyond 244 characters, the result of the evaluation entailed by = can only be to truncate the list, with possibly mysterious consequences.

The Stata Journal (2008)
8, Number 4, pp. 588–591

# Stata tip 71: The problem of split identity, or how to group dyads

Nicholas J. Cox
Department of Geography
Durham University
Durham City, UK
n.j.cox@durham.ac.uk

Many researchers in various disciplines deal with dyadic data, including several who would not use that term. Consider couples in pairings or relationships of any kind, including spouses, lovers, trading partners, countries at war, teams or individuals in sporting encounters, twins, parents and children, owners and pets, firms in mergers or acquisitions, and so on. Dyads can be symmetric or asymmetric; benign, malign, or neutral; exclusive (one partner can be in only one dyad with their unique partner) or not. See Kenny, Kashy, and Cook (2006) for an introduction to the area from a social- and behavioral-science viewpoint.

Behind this intriguing variety lies a basic question: How can we handle dyad identifiers in Stata datasets? A natural data structure reflects the split identity of dyads: each dyad necessarily has two identifiers that researchers will usually read into two variables. This is indeed natural but also often poses a problem that we will need to fix. Suppose that Joanna and Jennifer are twins and that Billy Bob and Peggy Sue are twins. We might have observations looking like this:

```
. list person twin
```

|     | person    | twin      |
|-----|-----------|-----------|
| 1.  | Joanna    | Jennifer  |
| 2.  | Jennifer  | Joanna    |
| 3.  | Billy Bob | Peggy Sue |
| 4.  | Peggy Sue | Billy Bob |

And other variables would record data on each person. So the other variables for observation 1 could record the height, weight, number of children, etc., for Joanna, and those variables for observation 2 could record the same data for Jennifer.

In general, the identifiers need not be real names but could be any convenient string or numeric tags. Problems will arise if identifiers are not consistent across the two identifier variables. So with string identifiers, capitalization and other spelling must be identical, and all leading and trailing spaces should be trimmed. See Herrin and Poen (2008) for detailed advice on cleaning up string variables. Also, in general, there is no assumption so far that each person occurs just once in the dataset. Frequently, we will have multiple observations on each person in panel datasets, or we will have similar setups for other dyadic data.

If your dataset contains many hundreds or thousands of observations, you need automated methods for handling identifiers. Editing by hand is clearly time consuming, tedious, and error prone.

How do we spell out to Stata that Joanna and Jennifer are a pairing? Here is a simple trick that leads readily to others. We can agree that Joanna and Jennifer have a joint identity, which is (alphabetically) Jennifer Joanna. So we just need to sort those identifiers by observation, or rowwise. This can be done as follows:

```
. generate first = cond(person < twin, person, twin)
. generate second = cond(person < twin, twin, person)
. list person twin first second
```

|     | person    | twin      | first     | second    |
| --- | --------- | --------- | --------- | --------- |
| 1.  | Joanna    | Jennifer  | Jennifer  | Joanna    |
| 2.  | Jennifer  | Joanna    | Jennifer  | Joanna    |
| 3.  | Billy Bob | Peggy Sue | Billy Bob | Peggy Sue |
| 4.  | Peggy Sue | Billy Bob | Billy Bob | Peggy Sue |

This is all breathtakingly simple and obvious once you see the trick. You need to see that inequalities can be resolved for string arguments as well as for numeric arguments, and equally that `cond()` will readily produce string results if instructed.

Let us go through step by step. The `person < twin` string for Stata means that the value of `person` is less than the value of `twin`. For strings, *less than* means *earlier in alphanumeric order*. The precise order is that of `sort`, or of `char()`, not that of your dictionary. So "a" is less than "b", but "B" is less than "a" because all uppercase letters are earlier in alphanumeric order than all lowercase letters. This precise order should only bite you if your identifiers are inconsistent, contrary to advice already given.

You might wonder quite how broad-minded Stata is in this territory. Do the functions `min()` and `max()` show the same generosity? No; `min("a", "b")` fails as a type mismatch.

The `cond()` function assigns results according to the answer to a question. See Kantor and Cox (2005) for a detailed introduction. If `person < twin`, then `first` takes on the values of `person` and `second` takes on the values of `twin`. If that is not true, then it is the other way around. Either way, `first` and `second` end up alphanumerically sorted.

I find it helpful to check through all the logical possibilities with an inequality, if only to reassure myself quickly that every possibility will produce the result I want. An inequality based on < will not be true if the operands satisfy > or if they satisfy ==. Here, if the names are the wrong way around, they get swapped in the results for `person` or `twin`, which is as intended. What is easier to overlook is the boundary case of equality. If the names are the same, they will also be swapped, but that makes no difference; no harm is done and no information is lost. In the example of twins, names being the same might seem unlikely, but perhaps someone just used surnames, and the surnames are identical. As usual, however, data-entry errors are another matter. In some other examples of dyads, there may be good reason for the names to be consistently the same; if so, the problem discussed here does not arise at all.

An advantage of using `cond()` is that exactly the same code applies to numeric identifiers. Conversely, if the identifiers were numeric, it would be fine to code

```
. generate first = min(person, twin)
. generate second = max(person, twin)
```

and a quick check shows that this works even if the identifiers are identical.

The problem is now all but solved. We can group observations for each dyad with

```
. by first second: command
```

and if we need a unique identifier for each dyad—it will come in useful sooner or later—we can get that by typing

```
. egen id = group(first second)
```

The `egen, group()` command yields identifiers that are integers 1 and above. Note also its handy `label` option. For more detail on such variables, see Cox (2007).

Tips for dyads should come in pairs, so here is another. This is for the case in which there are precisely two observations for each dyad. Again twins are a clear-cut example. Often we will want to compare each twin with the other, say, by calculating a difference. Then the height difference for each twin is *this twin's height* minus *the other twin's height*, or

```
. by first second: generate diffheight = height - height[3 - _n]
```

Or if we had calculated that identifier variable mentioned earlier, the height difference would be

```
. by id: generate diffheight = height - height[3 - _n]
```

Where does the `[3 - _n]` subscript come from? Recall that under the aegis of `by:`, `_n` is determined *within* groups defined by the *byvarlist*, here, the identifier variable `id`. For more on that, see Cox (2002) or, perhaps more conveniently, the *Speaking Stata* column in this issue (Cox and Longton 2008). So `_n` will be 1 or 2. If it is 1, then $3 - 1$ is 2, and if it is 2, then $3 - 2$ is 1.

If that seems too tricky to recall, there are more commonplace ways to do it:

```
. by id: generate diffheight =
> cond(_n == 1, height - height[2], height - height[1])
```

or even

```
. by id: generate diffheight = height - height[2] if _n == 1
. by id: replace diffheight = height - height[1] if _n == 2
```

# References

Cox, N. J. 2002. Speaking Stata: How to move step by: step. *Stata Journal* 2: 86–102.

———. 2007. Stata tip 52: Generating composite categorical variables. *Stata Journal* 7: 582–583.

Cox, N. J., and G. M. Longton. 2008. Speaking Stata: Distinct observations. *Stata Journal* 8: 557–568.

Herrin, J., and E. Poen. 2008. Stata tip 64: Cleaning up user-entered string variables. *Stata Journal* 8: 444–445.

Kantor, D., and N. J. Cox. 2005. Depending on conditions: A tutorial on the cond() function. *Stata Journal* 5: 413–420.

Kenny, D. A., D. A. Kashy, and W. L. Cook. 2006. *Dyadic Data Analysis*. New York: Guilford Press.

The Stata Journal (2008)
8, Number 4, pp. 592–593

# Stata tip 72: Using the Graph Recorder to create a pseudograph scheme

Kevin Crow
StataCorp
College Station, TX
kcrow@stata.com

Following the update of 25 February 2008, the Graph Editor can now record a series of edits, name the recording, and apply the edits from the recording to other graphs. You can apply the recorded edits from the Graph Editor or from the command line. The edits can be applied from the command line when a graph is created, when it is used from disk, or whenever it is the active graph. See *Graph Recorder* in `help graph editor` for creating and playing recordings in the Graph Editor. For applying edits from the command line, see `help graph play` and the option `play(`*recordingname*`)` in `help std_options` and `help graph use`.

In this tip, I focus on the use of the Graph Recorder to create a graph scheme. A graph scheme specifies the overall look of the graph. If you want to create your own look for your graphs, you will want to create a scheme file. There is a problem with scheme files, however, because unless you know exactly how you want your scheme to be set up, creating a scheme can be very time consuming.

A shortcut to creating a scheme file is saving a graph recording to disk and replaying that recording on your graphs by using the `play()` option of the `graph` command. Using the Graph Recorder to create your graph recording also allows you to tinker with your graph's look on the fly without having to edit a scheme file. Let's walk through an example.

Suppose you want to create several graphs for a report and you want those graphs to have a specific look. To create your recording, you first need to draw the first graph of the report. Try

```
. sysuse auto
(1978 automobile data)
. scatter mpg weight
```

Now that your graph is in the Graph window, you can right-click on the window and select **Start Graph Editor** to start the Graph Editor. Next start the Graph Recorder by clicking on the **Start Recording** button, ●, so that you save your changes to memory. Once the graph looks the way you want, you then click on the same button, ● (which now has the tool tip **End Recording**), to save your changes to a `.grec` file. By default, Stata saves `.grec` files to your PERSONAL/grec directory.

If you want to tinker with the graph during a recording, but you do not want the changes to be saved to the `.grec` file, click on the **Pause** button, ‖, to temporarily stop saving the changes. To unpause the Recorder, click on the **Pause** button, ‖, again.

Now that you have a recorder file saved to disk, you can type your next **graph** command in your do-file or from the Command window and apply your scheme to the graph with the **play()** option. For example,

```
. scatter mpg turn, play("test.grec") saving(test1, replace)
```

Also, if you have graphs already created and saved to disk, you can apply your recording to those graphs by using the **play()** option of **graph use**. For example,

```
. graph use "oldfile.gph", play("test.grec") saving("new_file", replace)
```

# Stata tip 73: append with care![1]

Christopher F. Baum
Department of Economics
Boston College
Chestnut Hill, MA
baum@bc.edu

The `append` command is a useful tool for data management. Most users are aware that they should be careful when appending datasets in which variable names differ; for instance, `PRICE` in one dataset with `price` in another will lead to both variables appearing in different columns of the combined dataset. But one perhaps lesser-known feature of `append` is worth noting. What if the *names* of the variables in the two datasets are the same, but their *data types* differ? If that is the case, then the order in which you combine the datasets may matter and can even lead to different retained contents in the combined dataset. This is particularly dangerous (as I recently learned!) when a variable is held as numeric in one dataset and string in another.

Let's illustrate this feature with `auto.dta`. You may know that the `foreign` variable is a 0/1 indicator variable (0 for domestic, 1 for foreign) with a value label. Let's create two datasets from `auto.dta`: the first with only domestic cars (`autodom.dta`) and the second with only foreign cars (`autofor.dta`). In the former dataset, we will leave the `foreign` variable alone. It is numeric and will be zero for all observations. In the second dataset, we create a string variable named `foreign`, containing `foreign` for each observation.

```
. sysuse auto
(1978 Automobile Data)
. drop if foreign
(22 observations deleted)
. save autodom
file autodom.dta saved
. sysuse auto
(1978 Automobile Data)
. drop if !foreign
(52 observations deleted)
. rename foreign nondom
. generate str foreign = "foreign" if nondom
. save autofor
file autofor.dta saved
```

Let's pretend that we are unaware that the same variable name, `foreign`, has been used to represent numeric content in one dataset and string content in the other, and let's use `append` to combine them. We use the domestic dataset and `append` the foreign dataset:

---

1. As of Stata 11, `append` will no longer append string variables to numeric variables or vice versa unless you use the `force` option, which should only be used with extreme care. This tip still applies to users of previous versions of Stata.—Ed.

```
. use autodom
(1978 Automobile Data)

. append using autofor
(note: foreign is str7 in using data but will be byte now)
(label origin already defined)

. describe foreign

              storage  display    value
variable name    type   format    label    variable label
──────────────────────────────────────────────────────────
foreign          byte   %8.0g     origin   Car type

. codebook foreign
──────────────────────────────────────────────────────────
foreign                                                Car type
──────────────────────────────────────────────────────────

             type:  numeric (byte)
            label:  origin

            range:  [0,0]                   units:  1
    unique values:  1                    missing .:  22/74

       tabulation:  Freq.  Numeric  Label
                       52        0  Domestic
                       22        .
```

Notice that **append** produced the following message:

```
(note: foreign is str7 in using data but will be byte now)
```

This message indicates that the **foreign** variable will be numeric in the combined dataset. The contents of the string variable in the using dataset have been lost, as you can see from the **codebook** output. Twenty-two cases are now classified as missing.

But quite a different outcome is forthcoming if we merely reverse the order of combining the datasets. It usually would not matter in what order we combined two or more datasets. After all, we could always use **sort** to place them in any desired order. But if we first **use** the foreign dataset and **append** the domestic dataset, we receive the following results:

```
. use autofor, clear
(1978 Automobile Data)

. append using autodom
(note: foreign is byte in using data but will be str7 now)
(label origin already defined)

. describe foreign

              storage  display    value
variable name    type   format    label    variable label
──────────────────────────────────────────────────────────
foreign          str7   %9s
```

```
. codebook foreign
```

---

foreign                                                                              (unlabeled)

---

```
              type:  string (str7)
      unique values:  1                              missing "":  52/74
         tabulation:  Freq.  Value
                         52   ""
                         22   "foreign"
```

Again we receive the fairly innocuous message:

```
(note: foreign is byte in using data but will be str7 now)
```

Unfortunately, this message may not call your attention to what has happened in the append. Because the data type of the first dataset rules, the string variable is unchanged, and the numeric values in the using dataset are discarded. As the codebook shows, the variable foreign is now missing for all domestic cars.

You may be used to the notion, with commands like merge, that the choice of master and using datasets matters. You may not be as well aware that append's results may also be sensitive to the order in which files are appended. You should always take heed of the messages that append produces when data types are altered. If the append step changes only the data type of a numeric variable to allow for larger contents (for instance, byte to long or float to double) or extends the length of a string variable to allow for longer strings, no harm is done. But append does not highlight the instances, such as those we have displayed above, where combining string and numeric variables with the same name causes the total loss of one or the other's contents. In reality, that type of data type alteration deserves a warning message, or perhaps even an error. Until and unless such changes are made to this built-in Stata command, append with care.

# Stata tip 74: firstonly, a new option for tab2

Roberto G. Gutierrez
StataCorp
College Station, TX
rgutierrez@stata.com

Peter A. Lachenbruch[1]
Department of Public Health
Oregon State University
Corvallis, OR
peter.lachenbruch@oregonstate.edu

In many research contexts, we have several categorical variables and correspondingly want to look at several contingency tables. The `tab2` command allows us to do this. Using the `auto` dataset (after categorizing `mpg` to `mpgcat` and `price` to `cost`), we have

```
. sysuse auto, clear
(1978 Automobile Data)
. quietly egen mpgcat = cut(mpg), at(0, 15, 25, 35, 45) label
. quietly egen cost = cut(price), at(0, 4000, 6000, 10000, 16000) label
. tab2 foreign mpgcat cost rep78, chi2
-> tabulation of foreign by mpgcat
```

|          |     | mpgcat |     |     |       |
|----------|-----|--------|-----|-----|-------|
| Car type | 0-  | 15-    | 25- | 35- | Total |
| Domestic | 7   | 37     | 8   | 0   | 52    |
| Foreign  | 1   | 10     | 8   | 3   | 22    |
| Total    | 8   | 47     | 16  | 3   | 74    |

```
          Pearson chi2(3) =  12.9821   Pr = 0.005
```

*(output omitted)*

```
-> tabulation of cost by rep78
```

|        |   | Repair Record 1978 |    |    |    |       |
|--------|---|--------------------|----|----|----|-------|
| cost   | 1 | 2                  | 3  | 4  | 5  | Total |
| 0-     | 0 | 1                  | 4  | 2  | 3  | 10    |
| 4000-  | 2 | 5                  | 17 | 8  | 6  | 38    |
| 6000-  | 0 | 1                  | 2  | 8  | 1  | 12    |
| 10000- | 0 | 1                  | 7  | 0  | 1  | 9     |
| Total  | 2 | 8                  | 30 | 18 | 11 | 69    |

```
          Pearson chi2(12) =  18.5482   Pr = 0.100
```

With 4 variables, we have 6 tables ($4 \times 3/2 = 6$). With more variables, the number of tables explodes quadratically. Even with 10 variables, we end up with 45 tables, which is likely to be more than really interests us. We may well want finer control.

Often there is one response variable of special interest. Our first focus may then be to relate that response to possible predictors. Suppose we wish to study if domestic or foreign cars differ on some variables. Thus we are interested in the three tables of `foreign` versus `mpgcat`, `cost`, and `rep78`. The `firstonly` option added to `tab2` in the update of 15 October 2008 allows us to get just the contingency table of the first-named variable versus the others.

---

1. Peter Lachenbruch's work was partially supported by a grant from the Cure JM Foundation.

```
. tab2 foreign mpgcat cost rep78, firstonly chi2
-> tabulation of foreign by mpgcat
```

|          |     | mpgcat |     |     |       |
|----------|-----|--------|-----|-----|-------|
| Car type | 0-  | 15-    | 25- | 35- | Total |
| Domestic | 7   | 37     | 8   | 0   | 52    |
| Foreign  | 1   | 10     | 8   | 3   | 22    |
| Total    | 8   | 47     | 16  | 3   | 74    |

```
          Pearson chi2(3) =  12.9821   Pr = 0.005
-> tabulation of foreign by cost
```

|          |     | cost  |       |        |       |
|----------|-----|-------|-------|--------|-------|
| Car type | 0-  | 4000- | 6000- | 10000- | Total |
| Domestic | 7   | 31    | 6     | 8      | 52    |
| Foreign  | 4   | 9     | 7     | 2      | 22    |
| Total    | 11  | 40    | 13    | 10     | 74    |

```
          Pearson chi2(3) =   5.3048   Pr = 0.151
-> tabulation of foreign by rep78
```

|          |   | Repair Record 1978 |    |    |    |       |
|----------|---|--------------------|----|----|----|-------|
| Car type | 1 | 2                  | 3  | 4  | 5  | Total |
| Domestic | 2 | 8                  | 27 | 9  | 2  | 48    |
| Foreign  | 0 | 0                  | 3  | 9  | 9  | 21    |
| Total    | 2 | 8                  | 30 | 18 | 11 | 69    |

```
          Pearson chi2(4) =  27.2640   Pr = 0.000
```

Here the number of tables is reduced from 6 to 3, a small change. However, for 10 variables (say, one response and nine predictors), the change is from 45 to 9.

This could have been programmed fairly easily with a foreach loop (see [P] **foreach**), but the new firstonly option makes life even a little easier.

The Stata Journal (2009)
**9**, Number 1, pp. 171–172

# Stata tip 75: Setting up Stata for a presentation

Kevin Crow
StataCorp
College Station, TX
kcrow@stata.com

If you plan to use Stata in a presentation, you might consider changing a few settings so that Stata is easy for your audience to view. How you set up Stata for presenting will depend on several factors like the size and layout of the room, the length of the Stata commands you will issue, the datasets you will use, the resolution of the projector, etc. Changing the settings and saving those settings as a custom preference before you present can save you time and frustration. Also having a custom layout preference allows you to restore your setup should something happen in the middle of your presentation.

How you manipulate Stata's settings is platform dependent. This article assumes you are using Windows. If you use Stata for Macintosh or Unix, the advice is the same but the manipulations are slightly different.

First, make Stata's windows fill the screen. The maximize button is in the top right-hand corner of Stata (the maximize button is in the same place for all windows in Stata). After maximizing Stata, you will also want to maximize the Results window.

Once Stata is maximized, you will probably want to move the Command window. For most room layouts, you will want the Command window at the top of Stata so that your audience can see the commands you are typing. You achieve this by changing your windowing preferences to allow docking. In Stata, select **Edit** > **Preferences** > **General Preferences...**, and then select the **Windowing** tab in the dialog box that appears. Make sure that the check box for **Enable ability to dock, undock, or tab windows** is checked, and then click on the **OK** button. Next double-click on the blue title bar of the Command window and drag the window to the top docking button. Once the Command window is docked on top, it is a good idea to go back to the *General Preferences* dialog box and uncheck the box you changed. Doing this will ensure that your Command window stays at the top of Stata and does not accidentally undock.

Depending on the projector resolution, you will probably want to change the font, font style, and font size of the Command window. To change the font settings of a window in Stata, right-click within the window and select **Font...**. The font you choose is up to you, but we recommend Courier New as a serif font or Lucida Console as a sans serif font. You will also want to change the font size (14 is a good starting size) and change the font style to bold. Finally, we recommend that you resize the Command window so that you can see two lines (with the font and font size changed, you might find that long Stata commands do not fit on one line).

Once the Command window is set, you now want to change the font and font size of the Results window. After you have the font and font size selected, be sure that the line size in the Results window is at least 80 characters long to prevent wrapping of output. You can check your line size by typing the following command in Stata.

```
. display c(linesize)
```

Another setting to consider changing is the color scheme of the Results window from the default black background scheme to the white background scheme. To do this, bring up the *General Preferences* dialog box and, in the **Results color** tab, change the **Color scheme** drop-down box to **White background**. Switching to this color scheme will help people in the audience who are color-blind.

Next change the font and font size of the Review and Variables windows. For the Variables window, you might want to resize the Name, Label, Type, or Format columns depending on your dataset. For example, if your dataset has long variable names but does not have variable labels, you would want to drag the Name column wider in the Variables window. If you plan to use the Viewer, Graph window, Do-file Editor, or Data Editor in your presentation, you will probably also want to resize the window and change the font and font size to make them easier to view.

You can do far more advanced Stata layouts by enabling some windowing preferences in Stata. For example, if you would like more room in the Results window, you might consider pinning the Review and Variables windows to the side of Stata. Again bring up the *General Preferences* dialog box in Stata and go to the **Windowing** tab. Check the box labeled **Enable ability to pin or unpin windows** and then close the dialog. You should now see a pin button in the blue title bars of the Review and Variables windows. Clicking on this button makes the windows a tab on the left side of Stata. To view the windows, simply click on the tab.

Finally, save your settings as a preference. In Stata, select **Edit > Preferences > Manage Preferences > Save Preferences > New Preferences Set...**. A dialog box will prompt you to name your preference. To load this saved preference, select **Edit > Preferences > Manage Preferences > Load Preferences**, and then select your preference listed in the menu.

The Stata Journal (2009)
**9**, Number 2, pp. 321–326

# Stata tip 76: Separating seasonal time series

Nicholas J. Cox
Department of Geography
Durham University
Durham, UK
n.j.cox@durham.ac.uk

Many researchers in various sciences deal with seasonally varying time series. The part rhythmic, part random character of much seasonal variation poses several graphical challenges for them. People usually want to see both the broad pattern and the fine structure of trends, seasonality, and any other components of variation. The very common practice of using just one plot versus date typically yields a saw-tooth or roller-coaster pattern as the seasons repeat. That method is often good for showing broad trends, but not so good for showing the details of seasonality. I reviewed several alternative graphical methods in a *Speaking Stata* column (Cox 2006). Here is yet another method, which is widely used in economics. Examples of this method can be found in Hylleberg (1986, 1992), Ghysels and Osborn (2001), and Franses and Paap (2004).

The main idea is remarkably simple: plot separate traces for each part of the year. Thus, for each series, there would be 2 traces for half-yearly data, 4 traces for quarterly data, 12 traces for monthly data, and so on. The idea seems unlikely to work well for finer subdivisions of the year, because there would be too many traces to compare. However, quarterly and monthly series in particular are so common in many fields that the idea deserves some exploration.

One of the examples in Franses and Paap (2004) concerns variations in an index of food and tobacco production for the United States for 1947–2000. I downloaded the data from http://people.few.eur.nl/paap/pbook.htm (this URL evidently supersedes those specified by Franses and Paap [2004, 12]) and named it `ftp`. For what follows, year and quarter variables are required, as well as a variable holding quarterly dates.

```
. egen year = seq(), from(1947) to(2000) block(4)
. egen quarter = seq(), to(4)
. gen date = yq(year, quarter)
. format date %tq
. tsset date
. gen growth = D1.ftp/ftp
```

Although a line plot is clearly possible, a scatterplot with marker labels is often worth trying first (figure 1). See an earlier tip by Cox (2005) for more examples.

```
. scatter growth year, ms(none) mla(quarter) mlabpos(0)
```

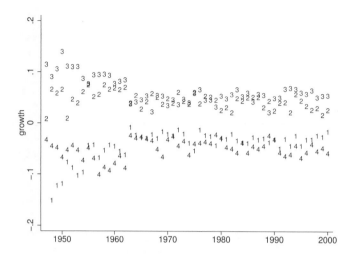

Figure 1. Year-on-year growth by quarter for food and tobacco production in the United States: separate series

Immediately, we see some intriguing features in the data. There seems to be a discontinuity in the early 1960s, which may reflect some change in the basis of calculating the index, rather than a structural shift in the economy or the climate. Note also that the style and the magnitude of seasonality change: look in detail at traces for quarters 1 and 4. No legend is needed for the graph, because the marker labels are self-explanatory. Compare this graph with the corresponding line plot given by Franses and Paap (2004, 15).

In contrast, only some of the same features are evident in more standard graphs. The traditional all-in-one line plot (figure 2) puts seasonality in context but is useless for studying detailed changes in its nature.

```
. tsline ftp
```

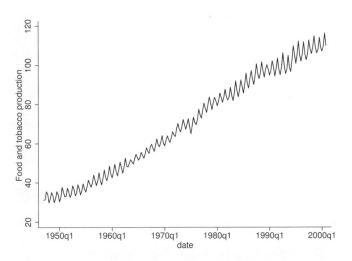

Figure 2. Quarterly food and tobacco production in the United States

The apparent discontinuity in the early 1960s is, however, clear in a plot of growth rate versus date (figure 3).

```
. tsline growth
```

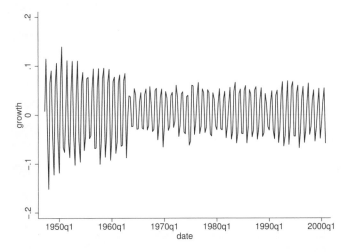

Figure 3. Year-on-year growth by quarter for food and tobacco production in the United States: combined series

An example with monthly data will push harder at the limits of this device. Grubb and Mason (2001) examined monthly data on air passengers in the United Kingdom for 1947–1999. The data can be found at http://people.bath.ac.uk/mascc/Grubb.TS; also see Chatfield (2004, 289–290). We will look at seasonality as expressed in monthly shares of annual totals (figure 4). The graph clearly shows how seasonality is steadily becoming more subdued.

```
. egen total = total(passengers), by(year)
. gen percent = 100 * passengers / total
. gen symbol = substr("1234567890ND", month, 1)
. scatter percent year, ms(none) mla(symbol) mlabpos(0) mlabsize(*.8) xtitle("")
> ytitle(% in each month) yla(5(5)15)
```

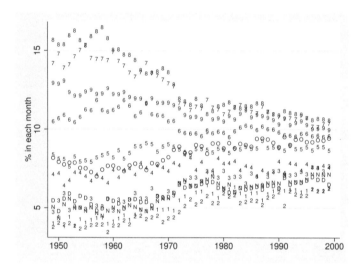

Figure 4. Monthly shares of UK air passengers, 1947–1999 (digits 1–9 indicate January–September; O, N, and D indicate October–December)

Because some users will undoubtedly want line plots, how is that to be done? The **separate** command is useful here: see Cox (2005), [D] **separate**, or the online help. Once we have separate variables, they can be used with the **line** command (figure 5).

```
. separate percent, by(month) veryshortlabel
. line percent1-percent12 year, xtitle("") ytitle(% in each month) yla(5(5)15)
> legend(pos(3) col(1))
```

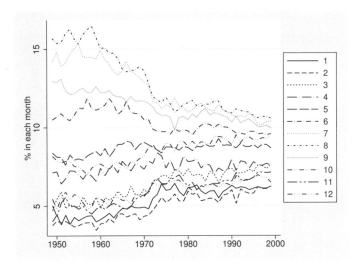

Figure 5. Monthly shares of UK air passengers, 1947–1999

You may think that the graph needs more work on the line patterns (and thus the legend), although perhaps now the scatterplot with marker labels seems a better possibility.

If graphs with 12 monthly traces seem too busy, one trick worth exploring is subdividing the year into two, three, or four parts and using separate panels in a by() option. Then each panel would have only six, four, or three traces.

# References

Chatfield, C. 2004. *The Analysis of Time Series: An Introduction.* 6th ed. Boca Raton, FL: Chapman & Hall/CRC.

Cox, N. J. 2005. Stata tip 27: Classifying data points on scatter plots. *Stata Journal* 5: 604–606.

———. 2006. Speaking Stata: Graphs for all seasons. *Stata Journal* 6: 397–419.

Franses, P. H., and R. Paap. 2004. *Periodic Time Series Models.* Oxford: Oxford University Press.

Ghysels, E., and D. R. Osborn. 2001. *The Econometric Analysis of Seasonal Time Series.* Cambridge: Cambridge University Press.

Grubb, H., and A. Mason. 2001. Long lead-time forecasting of UK air passengers by Holt–Winters methods with damped trend. *International Journal of Forecasting* 17: 71–82.

Hylleberg, S. 1986. *Seasonality in Regression*. Orlando, FL: Academic Press.

Hylleberg, S., ed. 1992. *Modelling Seasonality*. Oxford: Oxford University Press.

The Stata Journal (2009)
**9**, Number 3, pp. 497–498

# Stata tip 77: (Re)using macros in multiple do-files

Jeph Herrin
Yale School of Medicine
Yale University
New Haven, CT
jeph.herrin@yale.edu

Local and global macros provide an extremely useful way to define (for example) groups of values or variable names that can be (re)used in data management and analysis. For example, you may want to fit many different models using a given subset, or several different subsets, of independent variables. So you might define

```
. local indvars1 "var1 var2"
. local indvars2 "var1 var2 var3 var4"
```

Thereafter, subsequent models can be easily specified with reference to the same sets of variables:

```
. logit y `indvars1´
. regress y `indvars1´
. logit y `indvars2´
. regress y `indvars2´
```

However, users often like to define macros that can be referenced within several do-files. One way to have macros persist across do-files is to use globals, but globals should generally be reserved for macros that truly are wanted to be available in all contexts. An alternative is to borrow from other programming environments the use of "header" files, which contain preamble code that can be included at the top (or head) of each file of code to make common definitions.

For this purpose, Stata offers the `include` command, which is similar to `run` except that all local macro definitions are retained. See the manual entry [P] **include** for complete details. In the above example, we could have a file called `locals.do`:

———————————————————— begin `locals.do` ————————

```
local indvars1 "var1 var2"
local indvars2 "var1 var2 var3 var4"
```

———————————————————— end `locals.do` ————————

Then, in any file in which we would like to use these local macros, we can simply type

```
include locals.do
```

and thereafter refer to `indvars1`, `indvars2`, etc. If we subsequently want to modify the independent variables in our lists, we need only edit the `locals.do` file. The changes are automatically carried into other do-files that `include` that `locals.do` file.

This very useful feature can also be used to define directory paths and other programming data that we want to be available in local macros to all our do-files. For instance, a useful definition is

```
local today=string(date("`c(current_date)'","DMY"),"%tdCCYY.NN.DD")
```

which creates a string containing the current date in a lexically ordered format. Including this in a header file, and thence in all do-files, gives a standard way of adding a sortable date suffix to all saved files. You may `include` more than one file, as well as nesting `include`s, so that each project can `include` a `project.do` file that not only defines project-specific macros but also `include`s a `master.do`, which defines more general macros.

In summary, the use of `include` to call a file containing macro definitions allows one instance of those local (or global) macros to be made easily available to multiple do-files.

The Stata Journal (2009)
**9**, Number 3, pp. 499–503

# Stata tip 78: Going gray gracefully: Highlighting subsets and downplaying substrates

Nicholas J. Cox
Department of Geography
Durham University
Durham, UK
n.j.cox@durham.ac.uk

In graphics, as in life, going gray is often forced upon us, yet it is also occasionally a deliberate choice. Journals may enforce publication of your graphs in black and white whenever full-blown color is prohibited, or else prohibitively expensive. Even when allowed, color may prove problematic for various and quite different reasons, ranging from physiology and psychology to sociology and aesthetics. For example, many people are red–green color-blind, while the spectral or rainbow sequence from red to violet is not in fact perceived as a monotonic scale. Wilkinson (2005) and Ware (2008) give good introductions to the use of color in visualization. Fortner and Meyer (1997) give a more detailed discussion. Brewer (2005) gives many specific suggestions on color schemes, which are as appropriate for statistical graphics as they are for cartography.

Choosing differing shades of gray is also worth consideration for positive reasons. Expressing qualitative contrasts just with gray can be both effective and attractive. A previous Stata tip (Cox 2005) showed how distinct values on an ordered scale could be shown separately on scatterplots by markers of different gray-scale colors. This tip expands the theme with further examples. Naturally, black and white themselves qualify as extreme gray shades and may work very well. Here I will emphasize the use of intermediate shades. Let me underline that the examples here use the sj scheme. See [G-4] **schemes intro** for more information.

Consider the highlighting of subsets. You may want to show the distribution of a subset with the distribution of the complete set as context. Unwin, Theus, and Hofmann (2006); Chen, Härdle, and Unwin (2008); and Myatt and Johnson (2009) include several examples of this device for various kinds of graphs.

On a histogram with a frequency scale, this can be done by laying down the distribution of the complete set first and plotting the distribution of the subset on top. Display of the subset can never occlude the display of the complete set, because at most all the observations in any bin belong to the subset.

For example, after reading in a dataset from the U.S. National Longitudinal Study of Young Women in 1988,

```
. sysuse nlsw88
```

we may look at the wage distribution for college graduates compared with the complete set. Figure 1 shows such a histogram.

```
. twoway histogram wage, freq width(1) bcolor(gs14) blw(*.4) blcolor(black)
> || histogram wage if collgrad, yla(, ang(h)) xtitle(hourly wage (USD))
> ytitle(frequency) freq width(1) bcolor(gs6) blw(*.4) blcolor(black) legend(off)
```

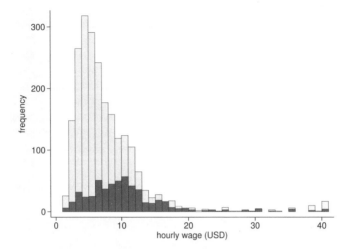

Figure 1. Wage distribution. College graduates are highlighted.

See [G-2] **graph twoway histogram** if you desire more detail on the command. The important detail here is spelling out that you want the same binning for comparability, as specified by width() and—if necessary—start(). Otherwise, the choices are matters of taste. For a written report, the legend is arguably dispensable, because what is being highlighted can be explained in the caption that you write within your word processor or text editor, as in this tip. For a talk, the need to make a graph self-explanatory might indicate otherwise.

The same distinction between complete set as backdrop and subset as highlight can be used in other plots. We will look at a scatterplot of wage against educational grade completed. Grade is discrete, but wage is not. We will do our own jittering of grade (only) by adding uniform noise beforehand, if only because that ensures consistency between graphs, and we plot wage on a logarithmic scale. Figure 2 shows a scatterplot with college graduates highlighted once again.

```
. generate grade2 = grade + .5 * (runiform() - .5)

. label var grade2 "`: var label grade´"

. scatter wage grade2, ms(Oh) mc(gs10) ysc(log)
> || scatter wage grade2 if collgrad, legend(off) ms(O) mc(gs2) ysc(log)
> ytitle(hourly wage (USD)) xla(0 4/18) yla(40 20 10 5 2, ang(h))
```

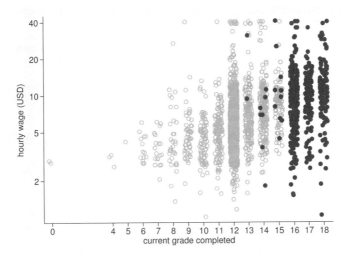

Figure 2. Wage and grade. College graduates are highlighted. Grade is jittered to give a better impression of variation.

Note that the ': ' construct inserts a variable label on the fly; see [P] **macro** for more details. Hollow and filled marker symbols, such as Oh and O, are helpfully complementary.

Sometimes we want to highlight an exploratory smooth or a fitted model prediction and correspondingly downplay the substrate of the data. With this dataset, noneconomists can join economists in being unsurprised at the great variability of wage within grade. All are likely to be much more interested in the average relationship. Nevertheless, suppressing the data on a graph would often be excessive, if not dishonest. Let us first smooth on a logarithmic scale using restricted cubic splines, experimenting only with default choices; [R] **mkspline** includes details and references. Note that the smooth is calculated for the unjittered grades.

```
. gen lnwage = ln(wage)
. mkspline spline = grade, cubic
. regress lnwage spline?
. predict smooth
```

Figure 3 shows a scatterplot with overlaid smooth. We have to do a little work to get *y*-axis labels in dollars. See Cox (2008) for further discussion. The logic, however, is easy: we just need to spell out which axis labels we want and where to put them. With this scheme, Stata automatically makes the scatterplot lighter than black, but it seems that we can fairly go further.

```
. scatter lnwage grade2, || mspline smooth grade, xla(0 4/18)
> legend(off) ytitle(hourly wage (USD))
> yla(`= ln(40)´ "40" `=ln(20)´ "20" `=ln(10)´ "10" `=ln(5)´ "5" `=ln(2)´ "2",
> ang(h))
```

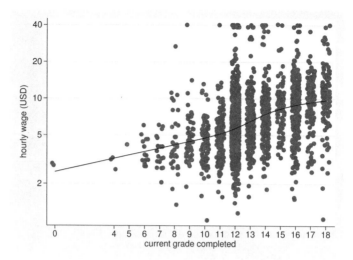

Figure 3. Restricted cubic spline smooth of wage versus grade on a logarithmic scale. Grade is jittered to give a better impression of variation in data.

Figure 4 shows the data points using a lighter gray and increases the width of the smooth. The result is likely to be closer to the researcher's message.

```
. scatter lnwage grade2, mcolor(gs10) || mspline smooth grade, lw(*3) lp(solid)
> xla(0 4/18) legend(off) ytitle(hourly wage (USD))
> yla(`= ln(40)´ "40" `=ln(20)´ "20" `=ln(10)´ "10" `=ln(5)´ "5" `=ln(2)´ "2",
> ang(h))
```

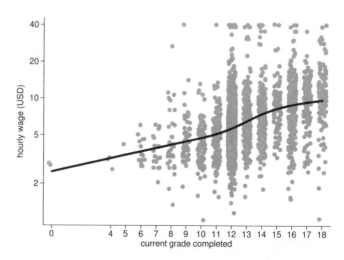

Figure 4. Restricted cubic spline smooth of wage on a logarithmic scale. Grade is jittered to give a better impression of variation in data. Note how the data are downplayed and the smooth highlighted compared with figure 3.

The ideas here can be taken in various further directions. Gray scale can be good for showing the scaffolding of the graph (axes, grids, and so forth) in a subdued but still discernible manner. Highlighting subsets can be extended to show three or more subsets. For example, to show a frequency-based histogram of three subsets, lay down the total distribution of all, followed by the total of two, followed by that of one, so that occlusion produces the desired effect. Alternatively, use `graph bar` or `graph hbar` to produce stacked or subdivided bars, introducing as much or as little histogram style as desired.

# References

Brewer, C. A. 2005. *Designing Better Maps: A Guide for GIS Users*. Redlands, CA: ESRI Press.

Chen, C., W. Härdle, and A. Unwin, eds. 2008. *Handbook of Data Visualization*. Berlin: Springer.

Cox, N. J. 2005. Stata tip 27: Classifying data points on scatter plots. *Stata Journal* 5: 604–606.

———. 2008. Stata tip 59: Plotting on any transformed scale. *Stata Journal* 8: 142–145.

Fortner, B., and T. E. Meyer. 1997. *Number by Colors: A Guide to Using Color to Understand Technical Data*. New York: Springer.

Myatt, G. J., and W. P. Johnson. 2009. *Making Sense of Data II: A Practical Guide to Data Visualization, Advanced Data Mining Methods, and Applications*. Hoboken, NJ: Wiley.

Unwin, A., M. Theus, and H. Hofmann. 2006. *Graphics of Large Datasets: Visualizing a Million*. New York: Springer.

Ware, C. 2008. *Visual Thinking for Design*. Burlington, MA: Morgan Kaufmann.

Wilkinson, L. 2005. *The Grammar of Graphics*. 2nd ed. New York: Springer.

The Stata Journal (2009)
**9**, Number 3, p. 504

# Stata tip 79: Optional arguments to options

Nicholas J. Cox
Department of Geography
Durham University
Durham, UK
n.j.cox@durham.ac.uk

Programmers occasionally would like an option for a program to come in two flavors: a simple or default option, with no arguments, and a more complicated but more flexible alternative, with arguments. For example, the simple option might call up a graph with programmer-chosen defaults, while the complicated option might pass graph options to the graph command in question, signaling variations from those defaults.

With a simple trick, you can implement two options that appear to the user to be this single option that is either simple or complicated. Following age-old programmer jargon, let us imagine an option that can be `foobar` or `foobar(arguments)`.

Step 1: Declare to `syntax` that there are two options, say, `foobar` and `FOOBAR2()`, and the latter is precisely, say, `FOOBAR2(string)`. The outburst of uppercase letters indicates to `syntax` that the latter can be abbreviated `foobar()`. You can also indicate names that can be abbreviated more, say, `FOObar` and `FOObar2(string)`.

Step 2: Process input within your program. For example,

```
if "`foobar'`foobar2'" != ""
```

is a test of whether either option has been called. If `foobar2()` has been called, then the local macro `foobar2` will be defined and can be treated further. If the argument to `foobar2()` might itself contain quotation marks, then compound double quotes, ` `" "' `, are in order.

Step 3: In documentation for the user, you need not mention the two options but may merely declare that the syntax is that an argument is optional, e.g., `foobar[ (string) ]`.

If curious users find out by looking at the code that the option with arguments is really `foobar2()`, no harm is done. They would be partway to working out, independently of this tip, how the optional options are coded. Browsing code and borrowing tricks that you want to use yourself remains one of the best ways to grow as a programmer.

The Stata Journal (2009)
**9**, Number 4, pp. 640–642

# Stata tip 80: Constructing a group variable with specified group sizes

Martin Weiss
Department of Economics
Tübingen University
Tübingen, Germany
martin.weiss@uni-tuebingen.de

Quite often, Stata users wish to construct a variable denoting group membership with different group sizes. This could be part of a simulation study in which a discrete variable with a certain distribution is required. Two cases can be distinguished: one in which the desired group sizes should be hit exactly and one in which group size should vary randomly around the desired proportion. In either case, group membership is to be determined randomly. This problem differs from grouping on the basis of one or more existing categories, which usually is best accomplished through egen, group() (see [D] **egen** and Cox [2007]).

The first goal can be achieved by obtaining uniformly distributed random numbers with Stata's runiform() function (see Buis [2007] for an expanded discussion of the usefulness of uniform random numbers) and sorting on these numbers. Because this uniform random variable is of no interest as such, we construct it as a temporary variable. The groups are subsequently determined in one fell swoop by conditioning on the position of an observation after the random sorting. The way to achieve this is to use an *expression*, as described in [U] **13 Functions and expressions**, featuring the system variable _n (see [U] **13.4 System variables (_variables)**), which denotes the running number of the observation. Say that you want groups with proportions 50%, 40%, and 10% in a sample of 10,000: whenever _n is smaller than or equal to 5,000, the expression inrange(_n,1,5000) evaluates to 1 and the other two evaluate to zero. Thus the entire sum evaluates to 1.

```
. clear
. set obs 10000
obs was 0, now 10000
. set seed 12345
. tempvar aux
. generate byte `aux´=runiform()
. sort `aux´
. generate byte group = inrange(_n,1,5000)*1+
>                       inrange(_n,5001,9000)*2+
>                       inrange(_n,9001,10000)*3
. tabulate group
```

|   group |    Freq. |  Percent |     Cum. |
|--------:|---------:|---------:|---------:|
|       1 |    5,000 |    50.00 |    50.00 |
|       2 |    4,000 |    40.00 |    90.00 |
|       3 |    1,000 |    10.00 |   100.00 |
|   Total |   10,000 |   100.00 |          |

The `set seed` command is used to make the draws from `runiform()`, and thus the division into groups, reproducible (for further information, see [R] **set seed**). This practice is also maintained in the examples below.

A feasible way to achieve the second goal is the use of nested `cond()` functions, as introduced in Kantor and Cox (2005):

```
. clear
. set obs 10000
obs was 0, now 10000
. set seed 12345
. generate byte group = cond(runiform()<0.5, 1, cond(runiform()<0.8, 2, 3))
. tabulate group
```

| group | Freq. | Percent | Cum. |
|-------|-------|---------|------|
| 1 | 4,947 | 49.47 | 49.47 |
| 2 | 4,048 | 40.48 | 89.95 |
| 3 | 1,005 | 10.05 | 100.00 |
| Total | 10,000 | 100.00 | |

The outer `cond()` function takes care of the 50% group, and the inner `cond()` function splits the remaining 50% in the ratio 4 to 1 so that the end result coincides with the user's intention. The two—separate—draws from `runiform()` are not related in any way.

This process can be shortened further with the help of the lesser-known `irecode()` function (see [D] **functions** for the full array of functions).

```
. clear
. set obs 10000
obs was 0, now 10000
. set seed 12345
. generate byte group = irecode(runiform(), 0, 0.5, 0.9, 1)
. tabulate group
```

| group | Freq. | Percent | Cum. |
|-------|-------|---------|------|
| 1 | 4,957 | 49.57 | 49.57 |
| 2 | 4,049 | 40.49 | 90.06 |
| 3 | 994 | 9.94 | 100.00 |
| Total | 10,000 | 100.00 | |

The `irecode()` function takes $n + 1$ arguments. The first argument is evaluated against the remaining $n$ arguments. If it is smaller than the second, "0" is returned; if it is larger than the second but smaller than the third, "1" is returned; and so on. Because the draws from `runiform()` must lie between 0 and 1, any draw will be larger than the second argument, and thus "0" will never be returned. Approximately 50% of the time, the draw will lie between the second and third argument, and the results will be "1", and so on. The desired proportions for the groups must be translated into differences between the arguments of the `irecode()` function: to get the 40% group, for instance, the third and fourth argument must be $0.9 - 0.5 = 0.4$ units apart.

As a final thought, consider the slight difference between the group sizes in the second and third example—although we did `set seed` to "12345" in both cases. As mentioned above, in the second example, two draws from the uniform distribution are conducted, one for each `cond()` function. The `irecode()` function in the third example, however, is fed one draw from `runiform()`. The function compares this one draw with the cutpoints provided by the user and assigns group membership accordingly.

## References

Buis, M. L. 2007. Stata tip 48: Discrete uses for uniform(). *Stata Journal* 7: 434–435.

Cox, N. J. 2007. Stata tip 52: Generating composite categorical variables. *Stata Journal* 7: 582–583.

Kantor, D., and N. J. Cox. 2005. Depending on conditions: A tutorial on the cond() function. *Stata Journal* 5: 413–420.

The Stata Journal (2009)
**9**, Number 4, pp. 643–647

# Stata tip 81: A table of graphs

Maarten L. Buis
Department of Sociology
Tübingen University
Tübingen, Germany
maarten.buis@uni-tuebingen.de

Martin Weiss
Department of Economics
Tübingen University
Tübingen, Germany
martin.weiss@uni-tuebingen.de

A useful option within the `graph` command is the `by()` option ([G-3] *by_option*). This option allows us to explore graphs across one or more additional categorical variables. This Stata tip will discuss how to fine-tune the display of such a graph when we want to compare graphs across two variables. Consider, for example, the graph created with the following set of commands and displayed in figure 1:

```
. sysuse auto
(1978 Automobile Data)

. fillin rep78 foreign

. twoway scatter price mpg, by(foreign rep78, cols(5) compact)
```

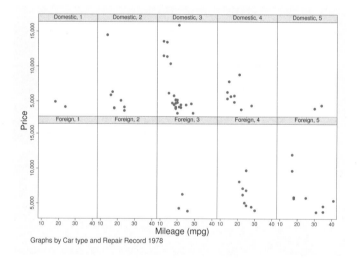

Figure 1. Comparing the relationship between price and mileage across repair status and car type (with a title for every subgraph)

There is a tabular look to this graph, with the rows representing the origin of the cars and the columns representing their repair status. This design helps us identify the relevant comparisons in a way similar to so-called trellis graphics (Cleveland 1993; Becker, Cleveland, and Shyu 1996) and lattice graphics (Sarkar 2008).

The first step in creating such a graph is to make sure that the columns and rows are properly aligned. We did that by using the `cols()` suboption of the `by()` option to make sure that there are as many columns as there are categories of `rep78`. We also used the `fillin` command ([D] **fillin**) to make sure that there is at least one observation for every combination of `rep78` and `foreign`, whereby those combinations that did not exist in the original data contain only missing values—here foreign cars with a repair status of poor (1) or fair (2). This has the effect of creating empty graphs for those nonexisting combinations. An alternative would be to use the `holes()` suboption within the `by()` option, which would leave the area otherwise occupied by a subgraph completely blank. Within the table-like logic of this graph, we like to use the empty graphs for combinations that in principle could occur but did not, while reserving those completely blank spaces for combinations that for logical reasons cannot occur. In our example, there is no reason why foreign cars could not have a poor or fair repair status, so we used the empty graphs. If we had compared graphs across gender and pregnancy status, we would instead have left a hole for the combination "pregnant men".

There is still a problem with this graph: the subgraph labels contain superfluous information and distract from the tabular design. Ideally, we would like to have titles at the top representing the columns and titles on the right representing the rows. To get the column titles, all we would need to do is keep only the top titles and suppress all the others. In the example below, this is done in the following way: the variables `rep78` and `foreign` are combined into one variable with the `egen` function `group()` ([D] **egen** and Cox [2007]). Then value labels that correspond to the column titles are assigned to the first five categories (`rep78` contains five categories, so in this graph there will be five columns). The remaining values of this new variable are assigned the value label " ". This will suppress the remaining titles. This example also adds titles to the column axis and the row axis. The resulting graph is shown in figure 2.

```
. egen group = group(foreign rep78)
(5 missing values generated)

. label define group 1  "Poor"
>                     2  "Fair"
>                     3  "Average"
>                     4  "Good"
>                     5  "Excellent"
>                     6  " "
>                     7  " "
>                     8  " "
>                     9  " "
>                    10  " "

. label value group group

. twoway scatter price mpg,
> by(group, cols(5)
>     r1title("Car type",
>             orientation(rvertical)
>             size(medsmall))
>     t1title("Repair status",
>             size(medsmall))
>     note("") compact)
```

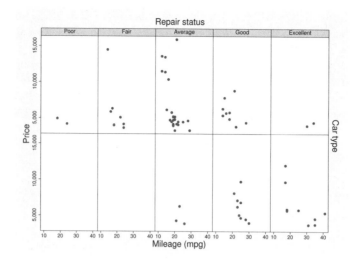

Figure 2. Comparing the relationship between price and mileage across repair status and car type (with column titles)

The row titles can now be added using the Graph Editor ([G-1] **graph editor** and Mitchell [2008, chapter 2]). The Graph Editor can be started within the graph window by going to the `file` menu and clicking on `Start Graph Editor`. Within the Graph Editor, the window of immediate interest is the `Object Browser` on the right-hand side of the Graph Editor. If we expand `plotregion1` (see figure 3) by clicking on the +, we can see that the characteristics of the individual subgraphs are identified by adding a [1] for the first (top left) subgraph, [2] for the second subgraph (to the right of the first subgraph), etc.

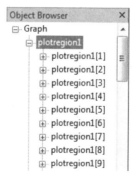

Figure 3. The object browser

In our example, we want to add a title to the right of the fifth and tenth subgraphs. A title to the right of a graph is an `r1title`, so within the `Object Browser`, we are looking for the elements `r1title[5]` for the top row title and `r1title[10]` for the bottom row title. This is specific to this example, because we wanted to create a graph with five columns and two rows. If, for example, we wanted to make a graph with three rows and three columns, we would be looking for r1title[3], r1title[6], and r1title[9], for the top, middle, and bottom row title, respectively.

When you double-click on r1title[5], you will get the dialog box shown in figure 4. In this dialog box, we can type the title. The remaining options govern how this row title will look. If we use the sj scheme ([G-4] **schemes intro**) for our graph, then the options shown in figure 4 will make sure that the row titles will look the same as the column titles. If you use another scheme, you can find out how to set those options by double-clicking on one of the column titles and looking in the resulting dialog box at how the options are set. The title for the bottom row is created in the same way by typing the title and setting the options for r1title[10]. The resulting graph is shown in figure 5.

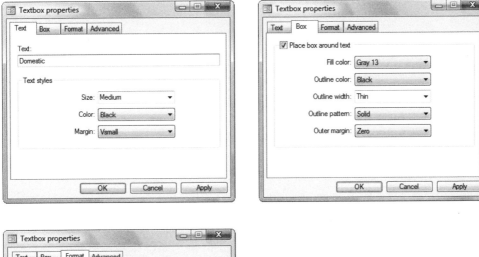

Figure 4. Properties specified in the Graph Editor for r1title[5]

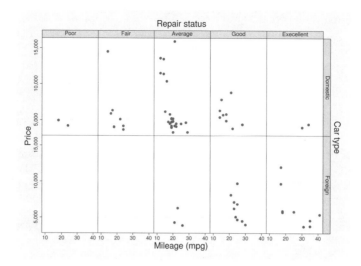

Figure 5. Comparing the relationship between price and mileage across repair status and car type (with column and row titles)

# References

Becker, R. A., W. S. Cleveland, and M.-J. Shyu. 1996. The visual design and control of trellis display. *Journal of Computational and Graphical Statistics* 5: 123–155.

Cleveland, W. S. 1993. *Visualizing Data*. Summit, NJ: Hobart.

Cox, N. J. 2007. Stata tip 52: Generating composite categorical variables. *Stata Journal* 7: 582–583.

Mitchell, M. N. 2008. *A Visual Guide to Stata Graphics*. 2nd ed. College Station, TX: Stata Press.

Sarkar, D. 2008. *Lattice: Multivariate Data Visualization with R*. New York: Springer.

The Stata Journal (2009)
**9**, Number 4, pp. 648–651

# Stata tip 82: Grounds for grids on graphs

Nicholas J. Cox
Department of Geography
Durham University
Durham, UK
n.j.cox@durham.ac.uk

Grids of horizontal or vertical lines within graphs were greatly used in the past, but more recently have been greatly disapproved. The need for grids as guidelines in preparing graphs has disappeared as computers have displaced people as graph constructors. Their use for precise look-up of particular values has diminished as more datasets become electronically accessible. To present tastes, past compendiums of graph types, such as Brinton (1914) and Karsten (1923), groan with the weight of heavy grids in example after example. The low point for grids came with the dismissal by Tufte (2001) of dark grid lines as "chartjunk", but he also emphasized that light grid lines could be helpful. The arguments of Tufte, Cleveland (1994), Kosslyn (2006), and others, and the flexibility of modern graphic technology, imply that grids should and can be subtle and subdued. Ideally, grids will be just noticeable so that they can be tuned in and out of attention by graph readers.

The grounds for grids are pragmatic and aesthetic—they can be useful and they can be pleasing. There is some room for disagreement on the first ground and much room on the second. The aim of this tip is not to change your mind on how you should prepare your own graphs, but more to underline some of the possibilities offered by Stata.

Stata provides support for grids, although you may easily not have noticed. For example, with the `auto.dta` dataset, Stata's default `s2color` graph scheme, and the canonical graph

```
. scatter mpg weight
```

subtle grid lines appear for `mpg` values 10, 20, 30, and 40. That also is true with the `sj` scheme used in the *Stata Journal*. These grid lines are associated with the corresponding axis labels. If you do not want them, you can turn them off with an option, such as `ylabel(, nogrid)`. Grid lines can also be associated with axis ticks: more usually, you would need to turn those on with an option, such as `ytick(15(10)35, grid)`—not that this is an especially good idea. Grid lines in the informal wider sense may also be added through options such as `yline()`. What goes for the $y$ axis also goes for the $x$ axis, at least as far as `twoway` is concerned.

To use grids effectively and tastefully, it is essential to be able to tune line width, color, and style, as well as horizontal or vertical position. Stata makes this easier by providing, at least in principle, `grid`, `major_grid`, and `minor_grid` *linestyle*s; see [G-4] *linestyle*. In practice, these three need not be distinct, depending precisely on the graph scheme in use. Independently of that, you can tweak grid lines as desired through standard line options.

As a first example, let us do what Arbuthnott (1710) did not do: plot his data on the ratio of males and females christened in London. The main statistical point for these data is that the average ratio is definitely not 1, as researchers would now typically flag with a significance test or confidence interval. The fuller history is enlightening and entertaining and can be found in many sources, including Stigler (1986) and Hacking (2006). Plotting these data was inspired by Friendly (2007). A data file is provided with the electronic media for this issue as `arbuthnott.dta`.

```
. use arbuthnott
```

The obvious reference lines for a time-series plot of the ratio are the line of equality, $y = 1$, and the line of the mean, $y = 1.066918$. Some experimenting indicates that `yline()` by itself produces rather stark lines, whereas `yline(, grid)` produces very subdued ones. One compromise is to specify `lcolor(gs12)`, thus making use of gray scale (Cox 2009). Adding these lines would make the graph appear a little busy given the grid lines provided by default to match the `ylabels` at 1, 1.05, 1.1, and 1.15, so we use `nogrid` to turn three of those off. Figure 1 is the result.

```
. line ratio year, yline(1.066918 1, lstyle(grid) lcolor(gs12))
> ylabel(, nogrid angle(h)) xtitle("")
```

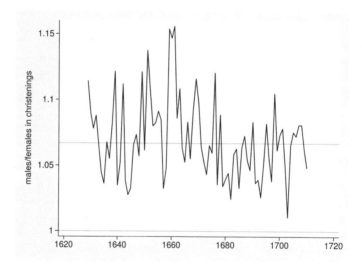

Figure 1. Arbuthnott's data on sex ratio in London christenings. The use of reference lines is shown in grid style.

Grids can be especially useful on scatterplots. Consider these data on city temperatures in the United States:

```
. sysuse citytemp
```

First, we create Celsius versions of each temperature variable:

```
. clonevar tempjanC = tempjan
. replace tempjanC = (5/9) * (tempjanC - 32)
. clonevar tempjulC = tempjuly
. replace tempjulC = (5/9) * (tempjulC - 32)
```

A natural reference line, which we make thicker than the default, is at the freezing point $0°C$ ($32°F$). We add more grid lines at other temperatures so that both axes are gridded. A major twist is that lighter and darker are inverted: the grid lines are white and the plot region and axes are set to a light gray, `gs14`. Figure 2 shows what this produces.

```
. scatter tempjulC tempjanC, ms(oh)
> xline(0, lstyle(grid) lcolor(white) lwidth(*1.5))
> xlabel(, grid glcolor(white)) ylabel(, angle(h) glcolor(white))
> plotregion(color(gs14)) graphregion(color(white)) xscale(lcolor(gs14))
> yscale(lcolor(gs14)) note({c 176}Celsius) ms(oh)
```

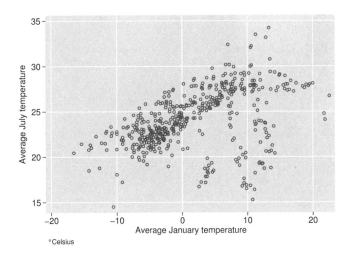

Figure 2. City temperature data for the United States. Grid lines are shown for both variables, and lighter and darker are inverted.

Such a look to graphs is in conscious imitation of a look very popular in the R community, particularly using its `ggplot2` package as devised and documented by Wickham (2009). Aside from that, the provision of grid lines is thus a natural way of emphasizing zeros or other key levels on one or both variables. Stata users sometimes want to move the axes so that they intersect at the origin (0, 0), just as they may well have been taught to do when young. Stata will not do that if either variable is ever negative, but clear grid lines provide a way to emphasize such levels while also ensuring that the data are not obscured by the axes or their associated labels, ticks, and titles.

# References

Arbuthnott, J. 1710. An argument for Divine Providence, taken from the constant regularity observed in the births of both sexes. *Philosophical Transactions of the Royal Society of London* 27: 186–190.

Brinton, W. C. 1914. *Graphic Methods for Presenting Facts*. New York: Engineering Magazine Company.

Cleveland, W. S. 1994. *The Elements of Graphing Data*. Rev. ed. Summit, NJ: Hobart.

Cox, N. J. 2009. Stata tip 78: Going gray gracefully: Highlighting subsets and downplaying substrates. *Stata Journal* 9: 499–503.

Friendly, M. 2007. A.-M. Guerry's *Moral Statistics of France*: Challenges for multivariable spatial analysis. *Statistical Science* 22: 368–399.

Hacking, I. 2006. *The Emergence of Probability: A Philosophical Study of Early Ideas about Probability, Induction and Statistical Inference*. Cambridge: Cambridge University Press.

Karsten, K. G. 1923. *Charts and Graphs: An Introduction to Graphic Methods in the Control and Analysis of Statistics*. New York: Prentice Hall.

Kosslyn, S. M. 2006. *Graph Design for the Eye and Mind*. New York: Oxford University Press.

Stigler, S. M. 1986. *The History of Statistics: The Measurement of Uncertainty before 1900*. Cambridge, MA: Belknap Press.

Tufte, E. R. 2001. *The Visual Display of Quantitative Information*. 2nd ed. Cheshire, CT: Graphics Press.

Wickham, H. 2009. *ggplot2: Elegant Graphics for Data Analysis*. New York: Springer.

The Stata Journal (2010)
**10**, Number 1, pp. 152–156

# Stata tip 83: Merging multilingual datasets

Devra L. Golbe
Hunter College, CUNY
New York
dgolbe@hunter.cuny.edu

merge is one of Stata's most important commands. Without specific instructions to the contrary, merge holds the master data file inviolate. Its variables are neither replaced nor updated. Variable and value labels are retained. All these properties and more are well documented. What is not so well documented is how merge interacts with Stata's multiple language support (label language), added in Stata 8.1 and described in Weesie (2005). In essence, Stata users must pay careful attention to which languages are defined and current when merging files.

The language feature is useful not only for multiple "real" languages (e.g., English and French) but also for using different sets of labels for different purposes, such as short labels ("Mines") and long ("Non-Metallic and Industrial Metal Mining") in one data file. merge may generate unexpected results if attention is not paid to the language definitions and, in particular, the current language in each file. Multilingual datasets to be merged should be defined with common languages and each should have the same language set as the current language.

I illustrate using auto.dta. Starting with autotech.dta, create a new dichotomous variable, guzzler, defined as mpg < 25, label the variable and its values in English (en) and French (fr), and save to a file called tech.dta. The tabulations below display variable and value labels:

```
. label language en

. tabulate guzzler
```

| Gas Guzzler | Freq. | Percent | Cum. |
|---|---|---|---|
| No | 19 | 25.68 | 25.68 |
| Yes | 55 | 74.32 | 100.00 |
| Total | 74 | 100.00 | |

```
. label language fr

. tabulate guzzler
```

| Verte | Freq. | Percent | Cum. |
|---|---|---|---|
| Oui | 19 | 25.68 | 25.68 |
| Non | 55 | 74.32 | 100.00 |
| Total | 74 | 100.00 | |

From `auto.dta`, create an English-labeled file, `origin.dta`. Now merge in `tech.dta` and tabulate `guzzler` and `foreign`:

```
. tabulate guzzler foreign
```

|       | Car type  |         |       |
|------:|----------:|--------:|------:|
| Verte | Domestic  | Foreign | Total |
|   Oui |         8 |      11 |    19 |
|   Non |        44 |      11 |    55 |
| Total |        52 |      22 |    74 |

`foreign` is labeled in English, but `guzzler` is labeled in French. How did that happen? The `label language` and `labelbook` commands can clarify:

```
. label language
Language for variable and value labels

    In this dataset, value and variable labels have been defined in only one
    language:  en
```

*(output omitted)*

```
. labelbook
```

```
value label origin
```

*(output omitted)*

```
definition
        0    Domestic
        1    Foreign
 variables:  foreign
```

```
value label yesno_en
```

*(output omitted)*

```
definition
        0    No
        1    Yes
 variables:
```

```
value label yesno_fr
```

*(output omitted)*

```
definition
        0    Oui
        1    Non
 variables:  guzzler
```

*(output omitted)*

On reflection, this is what we might have expected. The master dataset is inviolate in the sense that its language—English (and only English)—is preserved. The using dataset has two languages, but only the labels from the current language (French) are attached to the single language (English) in the master dataset.

We could, of course, redefine our languages and reattach the appropriate labels. However, if we plan to merge our master file with a multilingual dataset, a better strategy is to prepare the master file by defining the same two languages and then merge. It is crucial that the current languages are the same in both files. To illustrate, add French labels to origin.dta. Then consider what happens if the current languages differ. Suppose that the current language in the master file (origin.dta) is French, but the current language in the using file (tech.dta) is English. Then the labels from the current language in the using file (English) are attached in the current language of the master file (French), and the French labels (noncurrent) from the using file are not attached at all:

```
. label language fr
(fr already current language)

. tabulate guzzler foreign

     Gas |       Origine
 Guzzler |     USA     Autre  |     Total
---------+--------------------+----------
      No |       8        11  |        19
     Yes |      44        11  |        55
---------+--------------------+----------
   Total |      52        22  |        74

. label language en

. tabulate guzzler foreign

         |       Car type
 guzzler | Domestic   Foreign  |     Total
---------+--------------------+----------
       0 |       8        11  |        19
       1 |      44        11  |        55
---------+--------------------+----------
   Total |      52        22  |        74
```

Although we could of course correct the labels afterward, it is easier to make sure that the files are consistent before the merge. If we set the current language to English (or French) in both files before merging, we see that the labels are properly attached:

```
. label language en
(en already current language)

. tabulate guzzler foreign

     Gas |       Car type
 Guzzler | Domestic   Foreign  |     Total
---------+--------------------+----------
      No |       8        11  |        19
     Yes |      44        11  |        55
---------+--------------------+----------
   Total |      52        22  |        74

.   label language fr
```

```
. tabulate guzzler foreign
```

|        | Origine |       |       |
|--------|---------|-------|-------|
| Verte  | USA     | Autre | Total |
| Oui    | 8       | 11    | 19    |
| Non    | 44      | 11    | 55    |
| Total  | 52      | 22    | 74    |

Whether or not the labels are properly attached, merge does preserve them. But a little planning in advance will ensure that they are attached in the way you expect.

In passing, it is worth noting that the situation is more complex if there are no languages defined in the master file. In that case, issuing a label language statement changes the behavior of merge. First, let's see what happens if we ignore the language of the master file. Thus we start with the original autotech.dta and merge in the multilingual origin.dta. We see that two languages have been defined, and the labels are properly attached:

```
. webuse autotech, clear
(1978 Automobile Data)
. merge 1:1 make using origin, nogenerate
  (output omitted)
. tabulate foreign
```

| Car type | Freq. | Percent | Cum.   |
|----------|-------|---------|--------|
| Domestic | 52    | 70.27   | 70.27  |
| Foreign  | 22    | 29.73   | 100.00 |
| Total    | 74    | 100.00  |        |

```
. label language fr
. tabulate foreign
```

| Origine | Freq. | Percent | Cum.   |
|---------|-------|---------|--------|
| USA     | 52    | 70.27   | 70.27  |
| Autre   | 22    | 29.73   | 100.00 |
| Total   | 74    | 100.00  |        |

If before merging, however, we check that the master file's only language is the default, we get different results:

```
. webuse autotech, clear
(1978 Automobile Data)
. label language
Language for variable and value labels

    In this dataset, value and variable labels have been defined in only one
    language:  default
  (output omitted)
```

```
. merge 1:1 make using origin, nogenerate
```
*(output omitted)*
```
. label language
```
<u>Language for variable and value labels</u>
```
    In this dataset, value and variable labels have been defined in only one
    language:  default
```
*(output omitted)*

In this case, only the default language is defined. Tabulation of `foreign` indicates that labels are attached from only the current language in the using file. The reason is that issuing the `label language` command sets the characteristics that define the current language and all available languages. `merge` then declines to overwrite those characteristics with characteristics from the using dataset. If they are as yet undefined, however, those characteristics are taken from the using language.

# Reference

Weesie, J. 2005. Multilingual datasets. *Stata Journal* 5: 162–187.

# Stata tip 84: Summing missings

Nicholas J. Cox
Department of Geography
Durham University
Durham City, UK
n.j.cox@durham.ac.uk

Consider this pretend dataset and the result of a mundane `collapse`:

```
. list, sepby(g)
```

|      | g | y |
|------|---|---|
| 1.   | 1 | 1 |
| 2.   | 1 | 2 |
| 3.   | 1 | 3 |
| 4.   | 1 | 4 |
| 5.   | 2 | 5 |
| 6.   | 2 | 6 |
| 7.   | 2 | 7 |
| 8.   | 3 | 8 |
| 9.   | 3 | . |
| 10.  | 4 | . |

```
. collapse (sum) y, by(g)
. list
```

|      | g | y  |
|------|---|----|
| 1.   | 1 | 10 |
| 2.   | 2 | 18 |
| 3.   | 3 | 8  |
| 4.   | 4 | 0  |

Here we have a grouping variable, g, and a response, y. When we `collapse, by(g)` to produce a reduced dataset of the sums of y, many users are surprised at the results. Clearly $1+2+3+4 = 10$ and $5+6+7 = 18$, so no surprises there, but the other results need more comment.

When producing sums, Stata follows two specific rules:

1. Initialize the result at 0 and add pertinent numbers one by one until done. The result is the sum.

2. Ignore missings.

With these two rules, it should not surprise you that the sum $8 + .$ is reported as 8 (the missing is ignored) or that the sum of $.$ is reported as 0 (the missing is ignored, and we just see the result of initializing, namely, 0).

Nevertheless, many users see this reasoning as perverse. A counterargument runs like this: If we know that all values in a group are missing, then we know nothing about their sum, which should also be recorded as missing. Users with this view need to know how to get what they want. They should also want to know more about why Stata treats missings in this way.

Consider a related problem: What is the sum of an empty set? Stata's answer to the problem is best understood by extending the problem. Imagine combining that empty set with any nonempty set, say, 1, 2, 3. What is the sum of the combined set? The Stata answer is that as

$$\text{sum of empty set} + \text{sum of } 1, 2, 3 = \text{sum of combined set } 1, 2, 3$$

so also the sum of an empty set must be regarded as 0. Sums are defined not by what they are but by how they behave. Insisting that the sum of an empty set must be missing prevents you from ever combining that result helpfully with anything else.

This problem, which was dubbed "related", is really the same problem in Stata's mind. Because missings are ignored, it is as if they do not exist. Thus a set of values that is all missing is equivalent to an empty set.

If you look carefully, you will find this behavior elsewhere in Stata, so it should be less of a surprise. `summarize` (see [R] **summarize**) will also report the sum of missings as zero:

```
. sysuse auto, clear
(1978 Automobile Data)

. summmarize rep78 if rep78 == .

    Variable |       Obs        Mean    Std. Dev.       Min        Max
-------------+--------------------------------------------------------
       rep78 |         0

. return list

scalars:
                  r(N) =  0
              r(sum_w) =  0
                r(sum) =  0
```

The sum (and sum of weights) is not explicit in the output from `summarize`, but it is calculated on the side and put in r-class results.

`egen` (see [D] **egen**) functions `total()` and `rowtotal()` by default yield 0 for the total of missings. However, as from Stata 10.1, they now have a new `missing` option. With this option, if all arguments are missing, then the corresponding result will be set to missing.

Mata has an explicit variant on empty sets, e.g.,

```
. mata: sum(J(0,0,.))
  0
. mata: sum(J(0,0,42))
  0
```

Whenever this behavior does not meet your approval, you need a work-around. Returning to the first example, one strategy is to include in the `collapse` enough information to identify groups that were all missing. There are several ways to do this. Here is one:

```
. collapse (sum) y (min) min=y, by(g)
. list
```

|      | g | y  | min |
|------|---|----|-----|
| 1.   | 1 | 10 | 1   |
| 2.   | 2 | 18 | 5   |
| 3.   | 3 | 8  | 8   |
| 4.   | 4 | 0  | .   |

```
. replace y = . if missing(min)
(1 real change made, 1 to missing)
. list
```

|      | g | y  | min |
|------|---|----|-----|
| 1.   | 1 | 10 | 1   |
| 2.   | 2 | 18 | 5   |
| 3.   | 3 | 8  | 8   |
| 4.   | 4 | .  | .   |

The idea here is that the minimum is missing if and only if all values are missing. Using `missing(min)` as a test also catches any situation in which the smallest missing value is one of `.a`–`.z`.

Missings pose problems for any mathematical or statistical software, and it is difficult to design rules that match all users' intuitions. Indeed, some intuitions are changed by becoming accustomed to certain rules. Stata's philosophy of ignoring missings where possible contrasts with software in which missings are regarded as highly infectious, affecting whatever operations they enter. Here is not the place to debate the merits and demerits of different choices in detail but rather to underline that Stata does attempt to follow rules consistently, with the consequence that you need a work-around whenever you disagree with the outcome.

The Stata Journal (2010)
**10**, Number 1, pp. 160–163

# Stata tip 85: Looping over nonintegers

Nicholas J. Cox
Department of Geography
Durham University
Durham, UK
n.j.cox@durham.ac.uk

A loop over integers is the most common kind of loop over numbers in Stata programming, as indeed in programming generally. `forvalues`, `foreach`, and `while` may be used for such loops. See their manual entries in the *Programming Reference Manual* for more details if desired. Cox (2002) gives a basic tutorial on `forvalues` and `foreach`. In this tip, I will focus on `forvalues`, but the main message here applies also to the other constructs.

Sometimes users want to loop over nonintegers. The help for `forvalues` contains an example:

```
. forvalues x = 31.3 31.6 : 38 {
  2.        count if var1 < `x´ & var2 < `x´
  3.        summarize myvar if var1 < `x´
  4. }
```

It is perfectly legal to loop over such a list of numbers, because `forvalues` allows any arithmetic progression as lists, with either integer or noninteger constant difference between successive terms. However, such lists can cause problems. On grounds of precision, correctness, clarity, and ease of maintenance, the advice here is to use loops over integers whenever possible.

The precision problem is exactly that explained elsewhere (Cox 2006; Gould 2006; Linhart 2008). Stata necessarily works at machine level in binary, and so it does no calculations in decimal. Rather, it works with the best possible binary approximations of decimals and then converts to decimal digits for display. Not surprisingly, users often think in terms of decimals that they want to use in their calculations or display in their results. Commonly, the conversions required work well and are not detectable, but occasionally users can get surprising results. Here is a simple example:

```
. forvalues i = 0.0(0.05)0.15 {
  2. display `i´
  3. }
0
.05
.1
```

The user evidently expects `display` of 0, .05, .1, and .15 in turn, but the loop ends without displaying .15. Why is that? First, let us fix the loop by looping over integers and doing the noninteger arithmetic inside the loop. That is the most important trick for attacking this kind of problem.

```
. forvalues i = 0/3 {
  2. display `i´ * .05
  3. }
0
.05
.1
.15
```

Why did that work as desired, but not the previous loop? The default format for display is hiding from us the approximations that are being used, which are necessary because most multiples of 1/10 cannot be held as exact binary numbers. A format with many more decimal places reveals the problem:

```
. forvalues i = 0/3 {
  2. display %20.18f `i´ * .05
  3. }
0.000000000000000000
0.050000000000000003
0.100000000000000006
0.150000000000000022
```

The result 0.150000000000000022 is a smidgen too far as far as the first loop is concerned. Otherwise put, the loop terminates because Stata's approximation to $0.05 + 0.05 + 0.05$ is a smidgen more than its approximation to $0.15$:

```
. display %20.18f 0.05 + 0.05 + 0.05
0.150000000000000022
. display %20.18f 0.15
0.149999999999999994
```

A little more cryptic, but closer to the way that Stata actually works, is a display in hexadecimal:

```
. display %21x 0.05 + 0.05 + 0.05
+1.3333333333334X-003
. display %21x 0.15
+1.3333333333333X-003
```

The difference really is very small, but it is enough to undermine the intention behind the original loop.

Another trick that is sometimes useful to ensure desired results is the formatting of numerical values as desired. Leading zeros are often needed, and for that we just need to insist on an appropriate format:

```
. forvalues i = 1/20 {
  2. local j : display %02.0f `i´
  3. display "`j´"
  4. }
01
02
03
04
05
06
07
08
09
10
11
12
13
14
15
16
17
18
19
20
```

A subtlety to notice here is the pair of double quotes flagging to `display` that the macro j is to be treated as a string. Omitting the double quotes as in `display 'j'` would cause the formatting to be undone, because the default (numeric) display format would produce a display that started 1, 2, 3, and so forth. The syntax here for producing the local j is called an extended macro function and is documented in [P] **macro**.

If all that was desired was the display just seen, then the loop could be simplified to contain a single statement in its body, namely,

```
. display %02.0f `i´
```

However, knowing how to produce another macro with this technique has other benefits. One fairly common example is for cycling over filenames. Smart users know that it is a good idea to use a sequence of filenames such as `data01` through `data20`. Naming this way ensures that files will be listed by operating system commands in logical order; otherwise, the order would be `data1`, `data11`, and so forth. But those smart users then need Stata to reproduce the leading zero in any cycle over files. The loop above could easily be modified to solve that kind of problem by including a command such as

```
. use data`j´
```

Correctness, clarity, and ease of maintenance were also mentioned as advantages of looping over integers. Style preferences enter here, and programmers' experiences vary, but on balance fewer coding errors and clearer code overall seem likely to result from the approach here. Moreover, noninteger steps, such as .05 within the very first example, are rarely handed down from high as the only possibilities. There is a marked advantage to changing just a single constant rather than a series of values from problem to problem.

# References

Cox, N. J. 2002. Speaking Stata: How to face lists with fortitude. *Stata Journal* 2: 202–222.

———. 2006. Stata tip 33: Sweet sixteen: Hexadecimal formats and precision problems. *Stata Journal* 6: 282–283.

Gould, W. 2006. Mata Matters: Precision. *Stata Journal* 6: 550–560.

Linhart, J. M. 2008. Mata Matters: Overflow, underflow and the IEEE floating-point format. *Stata Journal* 8: 255–268.

The Stata Journal (2010)
**10**, Number 2, pp. 303–304

# Stata tip 86: The missing() function

Bill Rising
StataCorp
College Station, TX
brising@stata.com

Stata's treatment of missing numeric values in expressions is clear: numeric values behave like infinity. However, some care is needed whenever numeric missing values could appear in a relational expression. Typically, missing values are included or excluded explicitly by a segment of Stata code. But there is a better way to deal with missing values: the `missing()` function.

Suppose that we have a dataset containing the body mass index (BMI) and age for a sample of people. As happens in real-world datasets, some observations contain missing values for one or both of the variables. We would like to create an indicator variable that marks all the obese adults, meaning those who are at least 20 years old and have a BMI of at least 30. On the surface, the command we should type is

```
. generate obese_adult = age>=20 & bmi>=30
```

This will not yield the desired results, however, because the relational operators do not treat missing values as anything special—they are just considered to be very large numbers and so are still included. Hence, the values for `obese_adult` will be filled according to the following table:

|  | age<20 | age>=20, nonmissing | age missing |
|---|---|---|---|
| bmi<30 | 0 | 0 | 0 |
| bmi>=30, nonmissing | 0 | 1 | 1 |
| bmi missing | 0 | 1 | 1 |

The solution to this problem is to assign the value only for those observations that have no missing values. Because missing values are large, we could type

```
. generate obese_adult = age>=20 & bmi>=30 if age<. & bmi<.
```

Although correct, this is not the best solution. In particular, it is only readable to Stata users—other people would have no clue what `age<.` means.

A better way to work with missing values is to use the `missing()` function. Its syntax and behavior are simple: `missing(exp1[, exp2, ..., expN])` evaluates to 1 if any of the expressions is missing and 0 if none is missing. Thus we could rewrite our previous correct `generate` command as

```
. generate obese_adult = age>=20 & bmi>=30 if !missing(age,bmi)
```

This is much more readable.

Another advantage of using the `missing()` function is that it treats string and numeric variables in the same fashion. Now suppose that we have a dataset containing a string-valued identifier named `ssn` and a numeric identifier named `whodat`. If we want to mark all the observations for which at least one identifier is missing, we can do this quite easily:

```
. generate byte badid = missing(ssn,whodat)
```

Not only is this readable but it also does not require knowing the data type of the two variables.

One thing that can trip up users new to `missing()` is its arguments: they are a comma-separated list of expressions, not a varlist. If you are interested in using varlists, you can write your own ado-file:

```
*! version 1.0.0 February 18, 2010 @ 08:48:04
*! makes a comma-separated varlist
program define vcomma, rclass
version 11
        syntax [varlist]
    local varlist : subinstr local varlist " " ",", all
    return local varlist "`varlist'"
end
```

You can then give `vcomma` a varlist, and it will return the comma-separated list in `r(varlist)`. So, for example, if you were interested in finding any observation in a dataset that had missing values for any variable, you could do this with the following two lines:

```
vcomma
list if missing(`r(varlist)')
```

This is certainly much easier than using loops or a long `if` qualifier.

Working with missing values is necessary in most datasets. As you have learned in this tip, using the `missing()` function makes working with missing values less onerous.

The Stata Journal (2010)
**10**, Number 2, pp. 305–308

# Stata tip 87: Interpretation of interactions in nonlinear models

Maarten L. Buis
Department of Sociology
Tübingen University
Tübingen, Germany
maarten.buis@uni-tuebingen.de

When fitting a nonlinear model such as `logit` (see [R] **logit**) or `poisson` (see [R] **poisson**), we often have two options when it comes to interpreting the regression coefficients: compute some form of marginal effect or exponentiate the coefficients, which will give us an odds ratio or incidence-rate ratio. The marginal effect is an approximation of how much the dependent variable is expected to increase or decrease for a unit change in an explanatory variable; that is, the effect is presented on an additive scale. The exponentiated coefficients give the ratio by which the dependent variable changes for a unit change in an explanatory variable; that is, the effect is presented on a multiplicative scale. An extensive overview is given by Long and Freese (2006).

Sometimes, we are also interested in how the effect of one variable changes when another variable changes, called the interaction effect. Because there is more than one way in which we can define an effect in a nonlinear model, there must also be more than one way in which we can define an interaction effect. This tip deals with how to interpret these interaction effects when we want to present effects as odds ratios or incidence-rate ratios, which can be an attractive alternative to interpreting interactions effects in terms of marginal effects.

The motivation for this tip is many recent discussions on how to interpret interaction effects when we want to interpret them in terms of marginal effects (Ai and Norton 2003; Norton, Wang, and Ai 2004; Cornelißen and Sonderhof 2009). (A separate concern about interaction effects in nonlinear models that is often mentioned is the possible influence of unobserved heterogeneity on these estimates; for example, see Williams [2009]. But I will not deal with that potential problem here.) These authors point out a common mistake, interpreting the first derivative of the multiplicative term between two explanatory variables as the interaction effect. The problem with this is that we want the interaction effect between two variables ($x_1$ and $x_2$) to represent how much the effect of $x_1$ changes for a unit change in $x_2$. The effect of $x_1$, in the marginal effects metric, is the first derivative of the expected value of the dependent variable ($E[y]$) with respect to $x_1$, which is an approximation of how much $E[y]$ changes for a unit change in $x_1$. The interaction effect should thus be the cross partial derivative of $E[y]$ with respect to $x_1$ and $x_2$—that is, an approximation of how much the derivative of $E[y]$ with respect to $x_1$ changes for a unit change in $x_2$. In nonlinear models, this is typically different from the first derivative of $E[y]$ with respect to the multiplicative term $x_1 \times x_2$. This is where programs like `inteff` by Norton, Wang, and Ai (2004) and `inteff3` by Cornelißen and Sonderhof (2009) come in.

Fortunately, we can interpret interactions without referring to any additional program by presenting effects as multiplicative effects (for example, odds ratios, incidence-rate ratios, hazard ratios). However, the marginal effects and the multiplicative effects answer subtly different questions, and thus it is a good idea to have both tools in your toolbox.

The interpretation of results is best explained using an example. Here we study whether the effect of having a college degree (collgrad) on the odds of obtaining a "high" job (high_occ) differs between black and white women.

```
. sysuse nlsw88
(NLSW, 1988 extract)

. generate byte high_occ = occupation < 3 if occupation < .
(9 missing values generated)

. generate byte black = race == 2 if race < .

. drop if race == 3
(26 observations deleted)

. generate byte baseline = 1

. logit high_occ black##collgrad baseline, or noconstant nolog

Logistic regression                             Number of obs   =       2211
                                                Wald chi2(4)    =     504.62
Log likelihood = -1199.4399                     Prob > chi2     =     0.0000
```

| high_occ | Odds Ratio | Std. Err. | z | P>\|z\| | [95% Conf. Interval] | |
|---|---|---|---|---|---|---|
| 1.black | .4194072 | .0655069 | -5.56 | 0.000 | .3088072 | .5696188 |
| 1.collgrad | 2.465411 | .293568 | 7.58 | 0.000 | 1.952238 | 3.113478 |
| | | | | | | |
| black#<br>collgrad<br>1 1 | 1.479715 | .4132536 | 1.40 | 0.161 | .8559637 | 2.558003 |
| | | | | | | |
| baseline | .3220524 | .0215596 | -16.93 | 0.000 | .2824512 | .3672059 |

If we were to interpret these results in terms of marginal effects, we would typically look at the effect of the explanatory variables on the probability of attaining a high job. However, this example uses a logit model together with the or option, so the dependent variable is measured in the odds metric rather than in the probability metric. Odds have a bad reputation for being hard to understand, but they are just the expected number of people with a high job for every person with a low job. For example, the baseline odds—the odds of having a high job for white women without a college degree—is 0.32, meaning that within this category, we expect to find 0.32 women with a high job for every woman with a low job. The trick I have used to display the baseline odds is discussed in an earlier tip (Newson 2003). The odds ratio for collgrad is 2.47, which means that the odds of having a high job is 2.47 times higher for women with a college degree. There is also an interaction effect between collgrad and black, so this effect of having a college degree refers to white women. The effect of college degree for black women is 1.48 times that for white women. So the interaction effect tells how much the effect of collgrad differs between black and white women, but it does so in multiplicative terms. The results also show that this interaction is not significant.

This example points to the difference between marginal effects and multiplicative effects. Now we can compute the marginal effect as the difference between the expected odds of women with and without a college degree, rather than as the derivative of the expected odds with respect to collgrad. The reason for computing the marginal effect as a difference is that collgrad is a categorical variable, so this discrete difference corresponds more closely with what would actually be observed. Although it is a slight abuse of terminology, I will continue to call it the marginal effect.

The margins command below shows the odds of attaining a high job for every combination of black and collgrad. The odds of attaining a high job for white women without a college degree is 0.32, while the odds for white women with a college degree is 0.79. The marginal effect of collgrad for white women is thus 0.47. The marginal effect of collgrad for black women is only 0.36. The marginal effect of collgrad is thus larger for white women than for black women, while the multiplicative effect of collgrad is larger for black women than for white women.

```
. margins, over(black collgrad) expression(exp(xb())) post
Predictive margins                              Number of obs    =        2211
Model VCE      : OIM

Expression     : exp(xb())
over           : black collgrad
```

|                    | Margin    | Delta-method Std. Err. | z     | P>\|z\| | [95% Conf. Interval] |          |
|--------------------|-----------|------------------------|-------|---------|----------------------|----------|
| black# collgrad    |           |                        |       |         |                      |          |
| 0 0                | .3220524  | .0215596               | 14.94 | 0.000   | .2797964             | .3643084 |
| 0 1                | .7939914  | .078188                | 10.15 | 0.000   | .6407457             | .9472371 |
| 1 0                | .1350711  | .0190606               | 7.09  | 0.000   | .097713              | .1724292 |
| 1 1                | .4927536  | .1032487               | 4.77  | 0.000   | .29039               | .6951173 |

```
. lincom 0.black#1.collgrad - 0.black#0.collgrad
 ( 1)  - 0bn.black#0bn.collgrad + 0bn.black#1.collgrad = 0
```

|     | Coef.   | Std. Err. | z    | P>\|z\| | [95% Conf. Interval] |          |
|-----|---------|-----------|------|---------|----------------------|----------|
| (1) | .471939 | .081106   | 5.82 | 0.000   | .3129742             | .6309038 |

```
. lincom 1.black#1.collgrad - 1.black#0.collgrad
 ( 1)  - 1.black#0bn.collgrad + 1.black#1.collgrad = 0
```

|     | Coef.     | Std. Err. | z    | P>\|z\| | [95% Conf. Interval] |          |
|-----|-----------|-----------|------|---------|----------------------|----------|
| (1) | .3576825  | .1049933  | 3.41 | 0.001   | .1518994             | .5634656 |

The reason for this difference is that the multiplicative effects are relative to the baseline odds in their own category. In this example, these baseline odds differ substantially between black and white women: for white women without a college degree, we expect to find 0.32 women with a high job for every woman with a low job, while for black women without a college degree, we expect to find only 0.14 women with a high job for every woman with a low job. So even though the increase in odds as a result of getting a college degree is higher for white women than for black women, this increase as a percentage of the baseline value is less for white women than for black women. The multiplicative effects control in this way for differences between the groups in baseline odds. However, notice that marginal and multiplicative effects are both accurate representations of the effect of a college degree. Which effect one wants to report depends on the substantive question, whether or not one wants to control for differences in the baseline odds.

The example here is relatively simple with only binary variables and no controlling variables. However, the basic argument still holds when using continuous variables and when controlling variables are added. Moreover, the argument is not limited to results obtained from `logit`. It applies to all forms of multiplicative effects, and so, for example, to odds ratios from other models such as `ologit` (see [R] **ologit**) and `glogit` ([R] **glogit**); relative-risk ratios ([R] **mlogit**); incidence-rate ratios (for example, [R] **poisson**, [R] **nbreg**, and [R] **zip**); or hazard ratios (for example, [ST] **streg** and [R] **cloglog**).

# Acknowledgment

I thank Richard Williams for helpful comments.

# References

Ai, C., and E. C. Norton. 2003. Interaction terms in logit and probit models. *Economics Letters* 80: 123–129.

Cornelißen, T., and K. Sonderhof. 2009. Partial effects in probit and logit models with a triple dummy-variable interaction term. *Stata Journal* 9: 571–583.

Long, J. S., and J. Freese. 2006. *Regression Models for Categorical Dependent Variables Using Stata*. 2nd ed. College Station, TX: Stata Press.

Newson, R. 2003. Stata tip 1: The eform() option of regress. *Stata Journal* 3: 445.

Norton, E. C., H. Wang, and C. Ai. 2004. Computing interaction effects and standard errors in logit and probit models. *Stata Journal* 4: 154–167.

Williams, R. 2009. Using heterogenous choice models to compare logit and probit coefficients across groups. *Sociological Methods and Research* 37: 531–559.

The Stata Journal (2010)
**10**, Number 2, pp. 309–312

# Stata tip 88: Efficiently evaluating elasticities with the margins command

Christopher F. Baum
Department of Economics
Boston College
Chestnut Hill, MA
baum@bc.edu

The new `margins` command, available in Stata 11, greatly simplifies many postestimation computations. Discussions of this command have justifiably highlighted its capabilities to work with factor variables and expressions involving factor-variable operators, such as `c.x#c.x`. But you may find another aspect of `margins` very useful: its ability to compute quantities such as elasticities and semielasticities with a much simpler command syntax than that previously available.

As discussed in [R] **margins**, under the section titled *Expressing derivatives as elasticities*, the elasticity of $y$ with respect to $x$ is the proportional change in $y$ for a proportional change in $x$. A semielasticity such as `eydx` is the proportional change in $y$ from a unit change in $x$.

For instance, we might estimate a linear regression equation for per capita expenditures on gasoline in the U.S. market from a time-series dataset:

```
. use http://fmwww.bc.edu/ec-p/data/greene2008/gasolinemarket
. tsset Year, yearly
        time variable:  Year, 1953 to 2004
                delta:  1 year
. generate gaspc = GasExp/(Gasp*(Pop/1e6))
. generate lngaspc = log(gaspc)
. local allreg Income Gasp PNC PUC PPT PD PN PS
```

```
. regress gaspc `allreg' Year
```

| Source | SS | df | MS | | Number of obs = | 52 |
|---|---|---|---|---|---|---|
| | | | | | F( 9, 42) = | 530.82 |
| Model | 56.7083042 | 9 | 6.30092268 | | Prob > F = | 0.0000 |
| Residual | .49854905 | 42 | .011870215 | | R-squared = | 0.9913 |
| | | | | | Adj R-squared = | 0.9894 |
| Total | 57.2068532 | 51 | 1.121703 | | Root MSE = | .10895 |

| gaspc | Coef. | Std. Err. | t | P>|t| | [95% Conf. Interval] | |
|---|---|---|---|---|---|---|
| Income | .0002157 | .0000518 | 4.17 | 0.000 | .0001113 | .0003202 |
| Gasp | -.0110838 | .0039781 | -2.79 | 0.008 | -.019112 | -.0030557 |
| PNC | .0005774 | .0128441 | 0.04 | 0.964 | -.0253432 | .0264979 |
| PUC | -.0058746 | .0048703 | -1.21 | 0.234 | -.0157033 | .0039541 |
| PPT | .0069073 | .0048361 | 1.43 | 0.161 | -.0028524 | .016667 |
| PD | .0012289 | .0118818 | 0.10 | 0.918 | -.0227495 | .0252072 |
| PN | .0126905 | .012598 | 1.01 | 0.320 | -.0127333 | .0381142 |
| PS | -.0280278 | .0079962 | -3.51 | 0.001 | -.0441649 | -.0118907 |
| Year | .0725037 | .0141828 | 5.11 | 0.000 | .0438816 | .1011257 |
| _cons | -140.4213 | 27.19985 | -5.16 | 0.000 | -195.3128 | -85.5298 |

```
. estimates store a
```

Let us say that we wish to calculate elasticities for the year 2004, using the regressors' values in that year. Using pre–Stata 11 syntax, we must store the regression estimates (as above) so that we can use the e-class command `mean` to evaluate the regressors' values in that year and store them in a row vector, x2004. We can then restore the regression estimates and use `mfx compute, eyex` to compute the elasticities, with its `at()` option specifying the point in the regressors' space at which they are to be evaluated:

```
. // elasticities at means: compute at t=2004 the old way (with mfx)
. quietly mean `allreg' Year if Year==2004

. matrix x2004 = e(b)

. estimates restore a
(results a are active now)

. mfx compute, eyex at(x2004)
Elasticities after regress
      y = Fitted values (predict)
        = 6.1726971
```

| variable | ey/ex | Std. Err. | z | P>|z| | [ 95% C.I. ] | | X |
|---|---|---|---|---|---|---|---|
| Income | .9476599 | .2263 | 4.19 | 0.000 | .504127 | 1.39119 | 27113 |
| Gasp | -.2224796 | .08093 | -2.75 | 0.006 | -.381102 | -.063857 | 123.901 |
| PNC | .0125245 | .2786 | 0.04 | 0.964 | -.533521 | .55857 | 133.9 |
| PUC | -.1268632 | .10488 | -1.21 | 0.226 | -.332432 | .078706 | 133.3 |
| PPT | .2339837 | .16441 | 1.42 | 0.155 | -.08826 | .556228 | 209.1 |
| PD | .0228545 | .22098 | 0.10 | 0.918 | -.410256 | .455965 | 114.8 |
| PN | .3540265 | .35281 | 1.00 | 0.316 | -.337474 | 1.04553 | 172.2 |
| PS | -1.011648 | .29332 | -3.45 | 0.001 | -1.58654 | -.436759 | 222.8 |
| Year | 23.53872 | 4.63929 | 5.07 | 0.000 | 14.4459 | 32.6316 | 2004 |

In contrast, what if we used `margins` to make this computation? We need not store the regression estimates, and one command does the job:

```
. // elasticities at means: compute at t=2004 the new way (with margins)
. margins if Year==2004, eyex(_all) at((means) _all)

Conditional marginal effects                     Number of obs    =          1
Model VCE     : OLS

Expression    : Linear prediction, predict()
ey/ex w.r.t.  : Income Gasp PNC PUC PPT PD PN PS Year
at            : Income          =     27113 (mean)
                Gasp            =   123.901 (mean)
                PNC             =     133.9 (mean)
                PUC             =     133.3 (mean)
                PPT             =     209.1 (mean)
                PD              =     114.8 (mean)
                PN              =     172.2 (mean)
                PS              =     222.8 (mean)
                Year            =      2004 (mean)
```

|  | ey/ex | Delta-method Std. Err. | z | P>\|z\| | [95% Conf. Interval] | |
|---|---|---|---|---|---|---|
| Income | .9476599 | .2262966 | 4.19 | 0.000 | .5041268 | 1.391193 |
| Gasp | -.2224796 | .0809311 | -2.75 | 0.006 | -.3811017 | -.0638575 |
| PNC | .0125245 | .2786 | 0.04 | 0.964 | -.5335214 | .5585704 |
| PUC | -.1268632 | .1048839 | -1.21 | 0.226 | -.332432 | .0787055 |
| PPT | .2339837 | .1644132 | 1.42 | 0.155 | -.0882602 | .5562276 |
| PD | .0228545 | .2209787 | 0.10 | 0.918 | -.4102557 | .4559647 |
| PN | .3540265 | .3528127 | 1.00 | 0.316 | -.3374737 | 1.045527 |
| PS | -1.011648 | .2933159 | -3.45 | 0.001 | -1.586536 | -.4367592 |
| Year | 23.53872 | 4.639292 | 5.07 | 0.000 | 14.44587 | 32.63156 |

We merely indicate, with an `if` *exp* clause, that we want marginal effects for the year 2004; that elasticities are to be computed for `_all` regressors; and that they are to be computed at the regressors' means (which are, of course, single values for that year).

We would find it just as straightforward to compute elasticities over the range of regressors' values, for instance, at deciles of their respective distributions. The resulting estimates could be stored in a Stata matrix:

```
. // calculate elasticities at deciles over full sample:
. forvalues i=10(10)90 {
  2.        quietly margins, eyex(_all) at((p`i') _all)
  3.        local rn "`rn' p`i'"
  4.        matrix nu = nullmat(nu) \ r(b)
  5. }
. matrix rownames nu = `rn'
```

```
. matrix list nu, format(%6.3f) ti("Elasticities")
nu[9,9]:  Elasticities
       Income    Gasp     PNC     PUC     PPT      PD      PN      PS    Year
p10     0.637  -0.063   0.009  -0.045   0.045   0.014   0.124  -0.196  43.934
p20     0.604  -0.057   0.008  -0.045   0.045   0.013   0.112  -0.192  38.315
p30     0.614  -0.051   0.007  -0.040   0.044   0.011   0.104  -0.187  31.412
p40     0.620  -0.052   0.006  -0.039   0.052   0.011   0.115  -0.214  27.265
p50     0.668  -0.098   0.008  -0.063   0.068   0.016   0.172  -0.334  26.614
p60     0.762  -0.149   0.011  -0.119   0.136   0.024   0.243  -0.547  26.836
p70     0.798  -0.149   0.012  -0.122   0.157   0.024   0.264  -0.650  25.338
p80     0.814  -0.150   0.013  -0.136   0.206   0.025   0.301  -0.791  25.029
p90     0.852  -0.150   0.013  -0.144   0.220   0.025   0.310  -0.854  23.397
```

We might also want to hold some regressors' values fixed and vary others across their decile ranges. This can be easily implemented with the `at()` option. For instance,

```
margins, eyex(_all) at((p25) Income (median) Gasp PNC PUC PPT PD PN PS)
```

would compute elasticities at the 25th percentile of `Income` and at the median values of other regressors.

The Stata Journal (2010)
**10**, Number 3, pp. 496–499

# Stata tip 89: Estimating means and percentiles following multiple imputation[1]

Peter A. Lachenbruch
Oregon State University
Corvallis, OR
peter.lachenbruch@oregonstate.edu

## 1 Introduction

In a statistical analysis, I usually want some basic descriptive statistics such as the mean, standard deviation, extremes, and percentiles. See, for example, Pagano and Gauvreau (2000). Stata conveniently provides these descriptive statistics with the `summarize` command's `detail` option. Alternatively, I can obtain percentiles with the `centile` command. For example, with `auto.dta`, we have

```
. sysuse auto
(1978 Automobile Data)

. summarize price, detail
```

```
                            Price

              Percentiles      Smallest
 1%            3291            3291
 5%            3748            3299
10%            3895            3667         Obs                 74
25%            4195            3748         Sum of Wgt.         74

50%            5006.5                       Mean           6165.257
                             Largest        Std. Dev.      2949.496
75%            6342           13466
90%           11385           13594         Variance        8699526
95%           13466           14500         Skewness       1.653434
99%           15906           15906         Kurtosis       4.819188
```

However, if I have missing values, the `summarize` command is not supported by `mi estimate` or by the user-written `mim` command (Royston 2004, 2005a,b, 2007; Royston, Carlin, and White 2009).

## 2 Finding means and percentiles when missing values are present

For a general multiple-imputation reference, see *Stata 11 Multiple-Imputation Reference Manual* (2009). By recognizing that a regression with no independent variables estimates the mean, I can use `mi estimate: regress` to get multiply imputed means. If I wish to get multiply imputed quantiles, I can use `mi estimate: qreg` or `mi estimate: sqreg` for this purpose.

---

1. Stata 11.2 and later will produce slightly different results because of improvements in the random-number generator.—Ed.

I now create a dataset with missing values of `price`:

```
. clonevar newprice = price
. set seed 19670221
. replace newprice = . if runiform() < .4
(32 real changes made, 32 to missing)
```

The following commands were generated from the multiple-imputation dialog box. I used 20 imputations. Before Stata 11, this could also be done with the user-written commands `ice` and `mim` (Royston 2004, 2005a,b, 2007; Royston, Carlin, and White 2009).

```
. mi set mlong
. mi register imputed newprice
(32 m=0 obs. now marked as incomplete)
. mi register regular mpg trunk weight length
. mi impute regress newprice, add(20) rseed(3252010)
Univariate imputation                      Imputations =        20
Linear regression                               added =        20
Imputed: m=1 through m=20                      updated =         0

                         |          Observations per m
                Variable |  complete  incomplete   imputed  |   total
                ---------+--------------------------------+--------
                newprice |        42          32        32  |      74

(complete + incomplete = total; imputed is the minimum across m
 of the number of filled in observations.)
. mi estimate: regress newprice
Multiple-imputation estimates              Imputations      =           20
Linear regression                          Number of obs    =           74
                                           Average RVI      =       1.3880
                                           Complete DF      =           73
                                           DF:       min    =        19.46
                                                     avg    =        19.46
DF adjustment:   Small sample                        max    =        19.46
                                           F(   0,      .) =
Within VCE type:          OLS              Prob > F         =            .

    newprice |      Coef.   Std. Err.      t    P>|t|     [95% Conf. Interval]
    ---------+----------------------------------------------------------------
       _cons |   5693.489   454.9877    12.51   0.000     4742.721    6644.258
```

From this output, we see that the estimated mean is 5,693 with a standard error of 455 (rounded up) compared with the complete data value of 6,165 with a standard error of 343 (also rounded up). However, we do not have estimates of quantiles. This could also have been done using `mi estimate: mean newprice` (the `mean` command is near the bottom of the estimation command list for `mi estimate`).

We can apply the same principle using `qreg`. For the 10th percentile, type

```
. mi estimate: qreg newprice, quantile(10)
Multiple-imputation estimates              Imputations      =        20
.1 Quantile regression                     Number of obs    =        74
                                           Average RVI      =    0.2901
                                           Complete DF      =        73
                                           DF:      min     =     48.05
                                                    avg     =     48.05
DF adjustment:    Small sample                      max     =     48.05
                                           F(    0,      .) =         .
                                           Prob > F         =         .
```

| newprice | Coef. | Std. Err. | t | P>|t| | [95% Conf. Interval] | |
|---|---|---|---|---|---|---|
| _cons | 3495.635 | 708.54 | 4.93 | 0.000 | 2071.058 | 4920.212 |

Compare the value of 3,496 with the value of 3,895 from the full data. We can use the simultaneous estimates command for the full set:

```
. mi estimate: sqreg newprice, quantiles(10 25 50 75 90) reps(20)
Multiple-imputation estimates              Imputations      =        20
Simultaneous quantile regression           Number of obs    =        74
                                           Average RVI      =    0.6085
                                           Complete DF      =        73
DF adjustment:    Small sample             DF:      min     =     23.19
                                                    avg     =     26.97
                                                    max     =     31.65
```

| newprice | Coef. | Std. Err. | t | P>|t| | [95% Conf. Interval] | |
|---|---|---|---|---|---|---|
| **q10** | | | | | | |
| _cons | 3495.635 | 533.5129 | 6.55 | 0.000 | 2408.434 | 4582.836 |
| **q25** | | | | | | |
| _cons | 4130.037 | 237.1932 | 17.41 | 0.000 | 3642.459 | 4617.614 |
| **q50** | | | | | | |
| _cons | 5200.238 | 441.294 | 11.78 | 0.000 | 4292.719 | 6107.757 |
| **q75** | | | | | | |
| _cons | 6620.232 | 778.8488 | 8.50 | 0.000 | 5025.49 | 8214.974 |
| **q90** | | | | | | |
| _cons | 8901.985 | 1417.022 | 6.28 | 0.000 | 5971.962 | 11832.01 |

# 3   Comments and cautions

The `qreg` command does not give the same result as the `centile` command when you have complete data. This is because the `centile` command uses one observation, while the `qreg` command uses a weighted combination of the observations. It will have somewhat shorter confidence intervals, but with large datasets, the difference will be small. A second caution is that comparing two medians can be tricky: the difference of two medians is not the median difference of the distributions. I have found it useful to use percentiles because there is a one-to-one relationship between percentiles if data are transformed. In our case, there is plentiful evidence that `price` is not normally distributed, so it would be good to look for a transformation and impute those values.

This method of using regression commands without an independent variable can provide estimates of quantities that otherwise would be difficult to obtain. For example, it is much faster than finding 20 imputed percentiles and then combining them with Rubin's rules, and it is much less onerous and prone to error.

## 4   Acknowledgment

This work was supported in part by a grant from the Cure JM Foundation.

## References

Pagano, M., and K. Gauvreau. 2000. *Principles of Biostatistics*. 2nd ed. Belmont, CA: Duxbury.

Royston, P. 2004. Multiple imputation of missing values. *Stata Journal* 4: 227–241.

———. 2005a. Multiple imputation of missing values: Update. *Stata Journal* 5: 188–201.

———. 2005b. Multiple imputation of missing values: Update of ice. *Stata Journal* 5: 527–536.

———. 2007. Multiple imputation of missing values: Further update of ice, with an emphasis on interval censoring. *Stata Journal* 7: 445–464.

Royston, P., J. B. Carlin, and I. R. White. 2009. Multiple imputation of missing values: New features for mim. *Stata Journal* 9: 252–264.

StataCorp. 2009. *Stata 11 Multiple-Imputation Reference Manual*. College Station, TX: Stata Press.

The Stata Journal (2010)
**10**, Number 3, pp. 500–502

# Stata tip 90: Displaying partial results

Martin Weiss
Department of Economics
Tübingen University
Tübingen, Germany
martin.weiss@uni-tuebingen.de

Stata provides several features that allow users to display only part of their results. If, for instance, you merely wanted to inspect the analysis of variance table returned by anova or the coefficients returned by regress, you could instruct Stata to omit other results:

```
. sysuse auto
(1978 Automobile Data)
. regress weight length price, notable
      Source |       SS       df       MS              Number of obs =      74
-------------+------------------------------           F(  2,    71) =  385.80
       Model | 40378658.3        2  20189329.2         Prob > F      =  0.0000
    Residual | 3715520.06       71  52331.2685         R-squared     =  0.9157
-------------+------------------------------           Adj R-squared =  0.9134
       Total | 44094178.4       73  604029.841         Root MSE      =  228.76
. regress weight length price, noheader
------------------------------------------------------------------------------
      weight |      Coef.   Std. Err.      t    P>|t|     [95% Conf. Interval]
-------------+----------------------------------------------------------------
      length |   30.60949   1.333171    22.96   0.000     27.95122    33.26776
       price |    .042138   .0100644     4.19   0.000     .0220702    .0622058
       _cons |  -2992.848   232.1722   -12.89   0.000    -3455.786    -2529.91
------------------------------------------------------------------------------
```

Other examples of this type can be found in the help files for xtivreg for its first-stage results and for xtmixed for its random-effects and fixed-effects table. Generally, to check whether Stata does provide such options, you would look for them under the heading *Reporting* in the respective help files.

If you want to further customize output to your own needs, you could use the estimates table command; see [R] **estimates table**. It is part of the comprehensive estimates suite of commands that save and manipulate estimation results in Stata. See [R] **estimates** or Baum (2006, sec. 4.4), where user-written alternatives are introduced as well.

estimates table can provide several benefits to the user. For one, you can restrict output to selected coefficients or equations with its keep() and drop() options.

```
. sysuse auto
(1978 Automobile Data)

. quietly regress weight length price trunk turn

. estimates table, keep(turn price)
```

| Variable | active |
|---|---|
| turn | 35.214901 |
| price | .04624804 |

The original output of the estimation command itself is suppressed with `quietly`; see [P] **quietly**. The `keep()` option also changes the order of the coefficients according to your wishes. Additionally, you can elect to have Stata display results in a specific format, for example, with fewer or more decimal places. The format can differ between the elements that you choose to put into the table. In the case shown below, the coefficients have three decimal places, while the standard error and the $p$-value have two decimal places:

```
. sysuse auto
(1978 Automobile Data)

. quietly regress weight length price trunk turn

. estimates table, keep(turn price) b(%9.3fc) se(%9.2fc) p(%9.2fc)
```

| Variable | active |
|---|---|
| turn | 35.215 |
|  | 11.65 |
|  | 0.00 |
| price | 0.046 |
|  | 0.01 |
|  | 0.00 |

legend: b/se/p

`estimates table` can also deal with models featuring multiple equations. If you want to omit the coefficients for `weight` and the constant from every equation of your `sureg` model, you could type

```
. sysuse auto
(1978 Automobile Data)

. qui sureg (price foreign weight length turn) (mpg foreign weight turn)

. estimates table, drop(weight _cons)
```

| Variable | active |
|---|---|
| price |  |
| foreign | 3320.6181 |
| length | −78.75447 |
| turn | −144.37952 |
| mpg |  |
| foreign | −2.0756325 |
| turn | −.23516574 |

If your interest rests in the entire first equation and the constant from the second equation, you would prepend coefficients with the equation names and separate the two with a colon. The names of equations and coefficients are more accessible in Stata 11 with the `coeflegend` option, which is accepted by most estimation commands.

```
. sureg, coeflegend noheader
```

|  | Coef. | Legend |
|---|---|---|
| **price** | | |
| foreign | 3320.618 | _b[price:foreign] |
| weight | 6.04491 | _b[price:weight] |
| length | -78.75447 | _b[price:length] |
| turn | -144.3795 | _b[price:turn] |
| _cons | 7450.657 | _b[price:_cons] |
| **mpg** | | |
| foreign | -2.075632 | _b[mpg:foreign] |
| weight | -.0055959 | _b[mpg:weight] |
| turn | -.2351657 | _b[mpg:turn] |
| _cons | 48.13492 | _b[mpg:_cons] |

```
. estimates table, keep(price: mpg:weight)
```

| Variable | active |
|---|---|
| **price** | |
| foreign | 3320.6181 |
| weight | 6.0449101 |
| length | -78.75447 |
| turn | -144.37952 |
| _cons | 7450.657 |
| **mpg** | |
| weight | -.00559588 |

See `help estimates table` to learn more about the syntax.

# Reference

Baum, C. F. 2006. *An Introduction to Modern Econometrics Using Stata.* College Station, TX: Stata Press.

The Stata Journal (2010)
**10**, Number 3, pp. 503–504

# Stata tip 91: Putting unabbreviated varlists into local macros

Nicholas J. Cox
Department of Geography
Durham University
Durham, UK
n.j.cox@durham.ac.uk

Within interactive sessions, do-files, or programs, Stata users often want to work with *varlists*, lists of variable names. For convenience, such lists may be stored in local macros. Local macros can be directly defined for later use, as in

```
. local myx "weight length displacement"
. regress mpg `myx´
```

However, users frequently want to put longer lists of names into local macros, spelled out one by one so that some later command can loop through the list defined by the macro. Such varlists might be indirectly defined in abbreviations using the wildcard characters * or ?. These characters can be used alone or can be combined to express ranges. For example, specifying * catches all variables, *TX* might define all variables for Texas, and *200? catches the years 2000–2009 used as suffixes.

In such cases, direct definition may not appeal for all the obvious reasons: it is tedious, time-consuming, and error-prone. It is also natural to wonder if there is a better method. You may already know that `foreach` (see [P] **foreach**) will take such wildcarded *varlist*s as arguments, which solves many problems.

Many users know that pushing an abbreviated *varlist* through `describe` or `ds` is one way to produce an unabbreviated *varlist*. Thus

```
. describe, varlist
```

is useful principally for its side effect of leaving all the variable names in `r(varlist)`. `ds` is typically used in a similar way, as is the user-written `findname` command (Cox 2010).

However, if the purpose is just to produce a local macro, the method of using `describe` or `ds` has some small but definite disadvantages. First, the output of each may not be desired, although it is easily suppressed with a `quietly` prefix. Second, the modus operandi of both `describe` and `ds` is to leave saved results as r-class results. Every now and again, users will be frustrated by this when they unwittingly overwrite r-class results that they wished to use again. Third, there is some inefficiency in using either command for this purpose, although you would usually have to work hard to measure it.

The solution here is to use the `unab` command; see [P] **unab**. `unab` has just one restricted role in life, but that role is the solution here. `unab` is billed as a programming command, but nothing stops it from being used interactively as a simple tool in data management. The simple examples

```
. unab myvars : *
. unab TX : *TX*
. unab twenty : *200?
```

show how a local macro, named at birth (here as `myvars`, `TX`, and `twenty`), is defined as the unabbreviated equivalent of each argument that follows a colon. Note that using wildcard characters, although common, is certainly not compulsory.

The word "unabbreviate" is undoubtedly ugly. The help and manual entry do also use the much simpler and more attractive word "expand", but the word "expand" was clearly not available as a command name, given its use for another purpose.

This tip skates over all the fine details of `unab`, and only now does it mention the sibling commands `tsunab` and `fvunab`, for use when you are using time-series operators and factor variables. For more information, see [P] **unab**.

# Reference

Cox, N. J. 2010. Speaking Stata: Finding variables. *Stata Journal* 10: 281–296.

The Stata Journal (2010)
**10**, Number 4, pp. 686–688

232

# Stata tip 92: Manual implementation of permutations and bootstraps

Lars Ängquist
Institute of Preventive Medicine
Copenhagen University Hospitals
Copenhagen, Denmark
la@ipm.regionh.dk

In mathematics, a permutation might be seen as a reordering of an ordered set of abstract elements (see, for example, Fraleigh [2002]),[1] whereas in data analysis—when facing empirical data—this concept may correspond to a reordering of an ordered set of observations. Vaguely speaking, in statistics and significance testing, this might be an interesting concept when simulating under a null hypothesis corresponding to, in some sense, a null association or an effect (most often an outcome) of one specific variable with respect to another one. Here one basically keeps the dataset constant except for the values, which are instead randomly permuted, corresponding to the core variable. Because all permutations are generally equally likely (at least if properly dealing with potential confounding) under the null hypothesis of no association, this is a way, through such simulations, of estimating the corresponding null distribution underlying, for instance, related $p$-values.

For similar reasons, one may apply the bootstrap simulation procedure. Here one does not reorder observations (or in general elements) but rather simulates from the empirical distribution based on this very set. In simulation terminology, the bootstrap and permutation procedures in this sense correspond to a uniformly random selection of values from the empirical distribution with and without replacement, respectively. For more information, see, for example, Manly (2007) for permutations, Davison and Hinkley (1997) for bootstraps, and Robert and Casella (2004) for stochastic simulation, in general.

In Stata, one may—given some assumed framework—use the commands `permute` and `bootstrap` to perform tasks related to permutation-based and bootstrap-based significance tests, respectively. Sometimes however, whether it arises as a need to be more specific or because one simply wants to keep more detailed control over the actual data manipulations, it might be favorable to perform some related manual labor at your computer keyboard. This tip is about the general structure of a solution for such a task.

**Permuting:** Assume that you have a variable of interest, `permvar`, that you want to permute in the sense noted above. Typing

---

[1]. The set of all possible reorderings (permutations) includes the permutation that actually leaves the order intact. This is called the identity permutation.

```
generate id=_n
generate double u=runiform()
sort u
local type: type permvar
generate `type´ upermvar=permvar[id]
```

in Stata will give you an additional column (`upermvar`) of permuted values. In the first command, a new variable, `id`, that corresponds to the current sort order is created.[2] In the second command, a column is generated with values uniformly distributed between 0 and 1. Because the values of `u` were randomly generated, sorting on `u` puts the observations in a random order. The next command saves the variable type of `permvar` in the local macro `type` so that the type can be applied to the new variable in the last command. The last command stores the permutation in the new variable `upermvar`: each new value is a value of `permvar` from a randomly selected observation. (The random selection is controlled by the `id` variable, which was put in a random order by the `sort` command.)

To reduce the risk of tied values with respect to the (inherently discrete) random draws, and moreover to further increase the, so to speak, randomness of the derived values, one might replace the code lines 2–3 with the following:

```
...
generate double u1=runiform()
generate double u2=runiform()
sort u1 u2
...
```

The randomness reference corresponds to the fact that computer-generated random numbers are random only to the extent permitted by the implementation of what is termed pseudorandom numbers (see, for instance, Knuth [1998]). To achieve reproducible results, one might take advantage of this pseudorandomness by explicitly stating a starting point, that is, a seed value, for the deterministic algorithm:

```
set seed 760130
```

The number must be a positive integer. For instance, this command might be used when assuring that different methods give equivalent results or, for example, with respect to estimated variances of certain derived estimates of interest, when comparing methods with respect to efficiency performance.[3]

---

2. In other words, this construction is based on the observation number indicator _n, which equals 1, 2, ..., N through the present observations (rows), where N is the number of observations in the dataset (generally reachable in a similar fashion through _N in Stata). Moreover, one approach to retaining a sort order, irrespective of the content of an executed program, is by taking advantage of the `sortpreserve` option (see `help program` or Newson [2004]).

3. You might use your personal birthdate as an easily remembered seed value. This is in fact used in the above case, though I am not revealing which date format I used; see `help dates and times`. I thank Claus Holst for this tip!

**Bootstrapping:** A related but slightly different variant of the above schedule might be used to derive a bootstrapped variable called `ubootsvar`. It is based on the empirical distribution formed or constituted by the present observations of the original variable `bootsvar`.

```
generate u=ceil(runiform()*_N)
generate ubootsvar=bootsvar[u]
```

Here the uniformly distributed values are not used to decide on a sort order (the underlying index values), but rather to directly constitute index values by making them be part of a uniformly distributed simulation of values on the integers 1, 2, ..., $N$. To achieve this, the so-called ceiling function, `ceil()`, is used. For more information on `runiform()`, see `help runiform` or Buis (2007)[4] (with respect to its use for simulations); further, `ceil()` and the related `floor()` function are described in Cox (2003).

Moreover, one might implement the above code structures into loops based on, for instance, `foreach` or `forvalues`. Under such circumstances, one might also take advantage of both usage of temporary variables (see `help tempvar`) and the specific matrix-oriented environment of Mata (see `help mata`) though the general structure described here might to some extent serve as a guideline or a template for such cases, as well. Once ready, strap your boots and let the permutation begin!

# References

Buis, M. L. 2007. Stata tip 48: Discrete uses for uniform(). *Stata Journal* 7: 434–435.

Cox, N. J. 2003. Stata tip 2: Building with floors and ceilings. *Stata Journal* 3: 446–447.

Davison, A. C., and D. V. Hinkley. 1997. *Bootstrap Methods and Their Application.* Cambridge: Cambridge University Press.

Fraleigh, J. B. 2002. *A First Course in Abstract Algebra.* 7th ed. Reading, MA: Addison–Wesley.

Knuth, D. E. 1998. *The Art of Computer Programming, Volume 2: Seminumerical Algorithms.* 3rd ed. Reading, MA: Addison–Wesley.

Manly, B. F. J. 2007. *Randomization, Bootstrap and Monte Carlo Methods in Biology.* 3rd ed. Boca Raton, FL: Chapman & Hall/CRC.

Newson, R. 2004. Stata tip 5: Ensuring programs preserve dataset sort order. *Stata Journal* 4: 94.

Robert, C. P., and G. Casella. 2004. *Monte Carlo Statistical Methods.* 2nd ed. New York: Springer.

---

4. The `uniform()` function was improved in Stata 11 and was renamed `runiform()`.

The Stata Journal (2010)
**10**, Number 4, pp. 689–690

# Stata tip 93: Handling multiple y axes on twoway graphs

Vince Wiggins
StataCorp
College Station, TX
vwiggins@stata.com

Sometimes users find it difficult to handle multiple $y$ axes on their `twoway` graphs. The main issue is controlling the side of the graph—left or right—where each axis is placed.

Here is a contrived example that exhibits the issue:

```
. sysuse auto
(1978 Automobile Data)
. collapse (mean) mpg trunk, by(length foreign)
. twoway bar    mpg length, yaxis(2) ||
        line trunk length, yaxis(1)
```

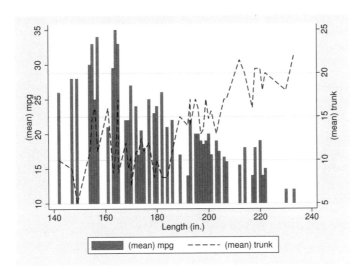

We might want `yaxis(1)` to be on the left of the graph and `yaxis(2)` to be on the right of the graph, but `twoway` insists on putting `yaxis(2)` on the left and `yaxis(1)` on the right. We could achieve what we want by reversing the order of the two plots, but the bars then occlude the lines, and who wants that?

It might be surprising, but the number assigned to an axis has nothing to do with its placement on the graph. `twoway` places the axes in the order in which it encounters them, with no consideration of their assigned number. How authoritarian! Consider `twoway`'s problem: when it sees `yaxis(2)`, it cannot be sure that it will ever see a `yaxis(1)`. Moreover, `twoway` will let you create more than two $y$ axes, and in that case it just stacks them up on the left of the graph like cordwood.

Do not worry. Although we may not like `twoway`'s rules, we can alter them. If we want any axes to appear in a different position, we just tell `twoway` to move them to the alternate (other) side of the graph using the `yscale(alt)` option. In this example, if we do not like the position of either $y$ axis, we will need to tell each of them to switch to the other side.

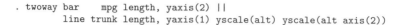

```
. twoway bar     mpg length, yaxis(2) ||
        line trunk length, yaxis(1) yscale(alt) yscale(alt axis(2))
```

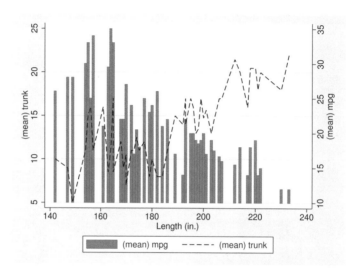

We typed `yscale(alt)` rather than the more explicit (but still valid) `yscale(alt axis(1))` because `axis(1)` is the default whenever we do not specify an axis. To alter the side where `axis(2)` appears, we had to be explicit about the axis number and type `yscale(alt axis(2))`.

If your axis is not where you want it, tell it to `alter` itself.[1]

# 1   Acknowledgment

I would like to thank *Stata Journal* editor Nicholas Cox for the initial adaptation of this tip from a Statalist posting, though Nick bears no responsibility for any remaining errors or puns.

---

1. The *Stata Journal* editors, against much stiff opposition, declare this to be the worst pun so far in the history of the *Stata Journal*.

The Stata Journal (2011)
11, Number 1, pp. 143–144

# Stata tip 94: Manipulation of prediction parameters for parametric survival regression models

Theresa Boswell
StataCorp
College Station, TX
tboswell@stata.com

Roberto G. Gutierrez
StataCorp
College Station, TX
rgutierrez@stata.com

After fitting a parametric survival regression model using `streg` (see [ST] `streg`), predicting the survival function for the fitted model is available with the `predict` command with the `surv` option. Some users may wish to alter the parameters used by `predict` to compute the survival function for a specific time frame or combination of covariate values.

Manipulation of the prediction parameters can be done directly by altering the variables that `predict` uses to calculate the survival function. However, it is good practice to create a copy of the variables before making any changes so that we can later return variables to their original forms.

This is best illustrated by an example. Using `cancer.dta` included with Stata, we can fit a simple Weibull model with one covariate, `age`:

```
. sysuse cancer
. streg age, dist(weibull)
```

Suppose that we want to obtain the predicted survival function for a specific time range and age value. The time variables used by `predict` to calculate the survival function are stored in variables `_t` and `_t0`, as established by `stset`. Before making any changes, we must first create a copy of these time variables and of our covariate `age`. We can use the `clonevar` command to create a copy. The advantage of using `clonevar` over `generate` is that `clonevar` creates an exact replica of each original variable, including its labels and other properties.

```
. clonevar age_orig = age
. clonevar t_orig = _t
. clonevar t0_orig = _t0
```

Now that we have a copy of the original variables, we are free to manipulate parameters. Let's assume that we want predictions of the survival function for individuals entering the study at age 75 over the time range [0,20]. To alter the time variables, we can use the `range` command to replace `_t` with an evenly spaced grid from 0 to 20:

```
. drop _t
. range _t 0 20
```

The `_t0` variable needs to be set to 0 (for obtaining unconditional survival), and `age` should be set to 75 for all observations:

```
. replace _t0 = 0
. replace age = 75
```

The prediction will now correspond to the survival function for an individual entering the study at age 75 over a time range of 0 to 20. The `predict` command with option `surv` will return the predicted survival function.

```
. predict s, surv
```

To view the predicted values, type

```
. list _t0 _t s
```

or you can graph the survival function by typing

```
. twoway line s _t
```

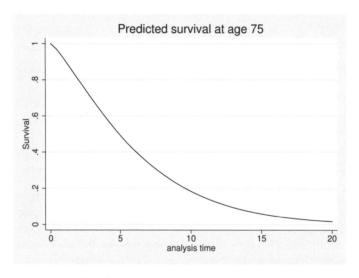

Figure 1. Predicted survival function

Now that we have the predicted values we want, it is prudent to replace all changed variables with their original forms. To do this, we will use the copies we created at the beginning of this example.

```
. replace age = age_orig
. replace _t = t_orig
. replace _t0 = t0_orig
```

There are many cases in which one may wish to manipulate the predicted survival function after `streg` and in which the steps in this tip can be followed to calculate the desired predictions.

The Stata Journal (2011)
**11**, Number 1, pp. 145–148

# Stata tip 95: Estimation of error covariances in a linear model

Nicholas J. Horton
Department of Mathematics and Statistics
Clark Science Center
Smith College
Northampton, MA
nhorton@smith.edu

## 1   Introduction

A recent review (Horton 2008) of the second edition of *Multilevel and Longitudinal Modeling Using Stata* (Rabe-Hesketh and Skrondal 2012) decried the lack of support in previous versions of Stata for models within the `xtmixed` command that directly estimate the variance–covariance matrix (akin to the REPEATED statement in SAS PROC MIXED). In this tip, I describe how support for these models is now available in Stata 11 (see also `help whatsnew10to11`) and demonstrate its use by replication of an analysis of a longitudinal dental study using an unstructured covariance matrix.

## 2   Model

I use the notation of Fitzmaurice, Laird, and Ware (2004, chap. 4 and 5) to specify linear models of the form $E(Y_i) = X_i\beta$, where $Y_i$ and $X_i$ denote the vector of responses and the matrix of covariates, respectively, for the $i$th subject, where $i = 1, \ldots, N$. Assume that each subject has up to $n$ observations on a common set of times. The response vector $Y_i$ is assumed to be multivariate normal with covariance given by $\Sigma_i(\theta)$, where $\theta$ is a vector of covariance parameters. If an unstructured covariance matrix is assumed, then there will be $n \times (n+1)/2$ covariance parameters. Restricted maximum-likelihood estimation is used.

## 3   Example

I consider data from an analysis of a study of dental growth, described on page 184 of Fitzmaurice, Laird, and Ware (2004). Measures of distances (in mm) were obtained on 27 subjects (11 girls and 16 boys) at ages 8, 10, 12, and 14 years.

### 3.1   Estimation in SAS

Below I give SAS code to fit a model with the mean response unconstrained over time (3 degrees of freedom) and main effect for gender as well as an unstructured working covariance matrix (10 parameters):

```
proc mixed data=one;
        class id time;
        model y = time female / s;
        repeated time / type=un subject=id r;
run;
```

This code generates the following output:

```
The Mixed Procedure
                  Model Information
Data Set                          WORK.ONE
Dependent Variable                y
Covariance Structure              Unstructured
Subject Effect                    id
Estimation Method                 REML
Residual Variance Method          None
Fixed Effects SE Method           Model-Based
Degrees of Freedom Method         Between-Within

              Dimensions
Covariance Parameters             10
Columns in X                       6
Columns in Z                       0
Subjects                          27
Max Obs Per Subject                4

            Estimated R Matrix for id 1
  Row      Col1        Col2        Col3        Col4
    1    5.3741      2.7887      3.8442      2.6242
    2    2.7887      4.2127      2.8832      3.1717
    3    3.8442      2.8832      6.4284      4.3024
    4    2.6242      3.1717      4.3024      5.3751

                  Solution for Fixed Effects
                          Standard
Effect        time   Estimate    Error    DF   t Value   Pr > |t|
Intercept            26.9258    0.5376    25    50.08     <.0001
time            8    -3.9074    0.4514    25    -8.66     <.0001
time           10    -2.9259    0.3466    25    -8.44     <.0001
time           12    -1.4444    0.3442    25    -4.20     0.0003
time           14         0         .     .        .          .
female               -2.0452    0.7361    25    -2.78     0.0102
```

## 3.2   Estimation in Stata

The equivalent model can now be fit in Stata 11:

```
. use http://www.math.smith.edu/labs/denttall
. xtmixed y ib14.time female, || id:, nocons residuals(un, t(time)) var
```

The `xtmixed` command yields the equivalent output:

```
Mixed-effects REML regression              Number of obs      =        108
Group variable: id                         Number of groups   =         27

                                           Obs per group: min =          4
                                                          avg =        4.0
                                                          max =          4

                                           Wald chi2(4)       =     101.50
Log restricted-likelihood = -212.4093      Prob > chi2        =     0.0000
```

| y | Coef. | Std. Err. | z | P>\|z\| | [95% Conf. Interval] | |
|---|---|---|---|---|---|---|
| **time** | | | | | | |
| 8 | -3.907407 | .4513647 | -8.66 | 0.000 | -4.792066 | -3.022749 |
| 10 | -2.925926 | .3466401 | -8.44 | 0.000 | -3.605328 | -2.246524 |
| 12 | -1.444444 | .3441962 | -4.20 | 0.000 | -2.119057 | -.7698322 |
| | | | | | | |
| female | -2.045172 | .736141 | -2.78 | 0.005 | -3.487982 | -.6023627 |
| _cons | 26.92581 | .5376092 | 50.08 | 0.000 | 25.87212 | 27.97951 |

| Random-effects Parameters | Estimate | Std. Err. | [95% Conf. Interval] | |
|---|---|---|---|---|
| **id:** (empty) | | | | |
| **Residual: Unstructured** | | | | |
| var(e8) | 5.374086 | 1.510892 | 3.097379 | 9.324271 |
| var(e10) | 4.21272 | 1.201038 | 2.409277 | 7.366114 |
| var(e12) | 6.428418 | 1.810989 | 3.700897 | 11.16609 |
| var(e14) | 5.375108 | 1.608682 | 2.989761 | 9.663575 |
| cov(e8,e10) | 2.788773 | 1.112924 | .6074823 | 4.970064 |
| cov(e8,e12) | 3.844272 | 1.392097 | 1.115811 | 6.572732 |
| cov(e8,e14) | 2.624241 | 1.207689 | .2572134 | 4.991268 |
| cov(e10,e12) | 2.883246 | 1.183372 | .5638802 | 5.202612 |
| cov(e10,e14) | 3.171762 | 1.153809 | .9103389 | 5.433186 |
| cov(e12,e14) | 4.302404 | 1.499388 | 1.363657 | 7.24115 |

```
LR test vs. linear regression:        chi2(9) =    54.59   Prob > chi2 = 0.0000
```

Note: The reported degrees of freedom assumes the null hypothesis is not on the boundary of the parameter space. If this is not true, then the reported test is conservative.

Several points are worth noting:

1. The default output from `xtmixed` provides estimates of variability as well as confidence intervals for the covariance parameter estimates.

2. Considerable flexibility regarding additional covariance structures is provided by the `residuals()` option (including exchangeable, autoregressive, and moving-average structures).

3. Specifying a `by()` variable within the `residuals()` option can allow separate estimation of error covariances by group (for example, in this setting, separate estimation of the structures for men and for women).

4. The `ib14` specification for the time factor variable facilitates changing the reference grouping to match the SAS defaults.

5. Dropping the `var` option will generate correlations (which may be more interpretable if the variances change over time).

For the dental example, we see that the estimated correlation is lowest between the observations that are farthest apart ($r = 0.49$) and generally higher for shorter intervals.

| corr(e8,e10) | .5861106 | .1306678 | .2743855 | .7863675 |
|---|---|---|---|---|
| corr(e8,e12) | .6540481 | .1129091 | .3761828 | .8239756 |
| corr(e8,e14) | .4882675 | .1518479 | .1420355 | .7280491 |
| corr(e10,e12) | .5540493 | .1370823 | .2322075 | .7665423 |
| corr(e10,e14) | .6665393 | .1115412 | .3894063 | .8330066 |
| corr(e12,e14) | .7319232 | .0930009 | .4931844 | .868134 |

# 4 Summary

Modeling the associations between observations on the same subject using mixed effects and an unstructured covariance matrix is a flexible and attractive alternative to a random-effects model with cluster–robust standard errors. This is particularly useful when the number of measurement occasions is relatively small, and measurements are taken at a common set of occasions for all subjects. The addition of support for this model within `xtmixed` in Stata 11 is a welcome development.

# 5 Acknowledgments

Thanks to Kristin MacDonald, Roberto Gutierrez, Garrett Fitzmaurice, and the anonymous reviewers for helpful comments on an earlier draft.

# References

Fitzmaurice, G. M., N. M. Laird, and J. H. Ware. 2004. *Applied Longitudinal Analysis*. Hoboken, NJ: Wiley.

Horton, N. J. 2008. Review of Multilevel and Longitudinal Modeling Using Stata, Second Edition, by Sophia Rabe-Hesketh and Anders Skrondal. *Stata Journal* 8: 579–582.

Rabe-Hesketh, S., and A. Skrondal. 2012. *Multilevel and Longitudinal Modeling Using Stata*. 3rd ed. College Station, TX: Stata Press.

The Stata Journal (2011)
**11**, Number 1, pp. 149–154

# Stata tip 96: Cube roots

Nicholas J. Cox
Department of Geography
Durham University
Durham, UK
n.j.cox@durham.ac.uk

## 1   Introduction

Plotting the graph of the cube function $x^3 = y$ underlines that it is single-valued and defined for arguments everywhere on the real line. So also is the inverse or cube root function $x = y^{1/3} = \sqrt[3]{y}$. In Stata, you can see a graph of the cube function by typing, say, `twoway function x^3, range(-5 5)` (figure 1). To see a graph of its inverse, imagine exchanging the axes. Naturally, you may want to supply other arguments to the `range()` option.

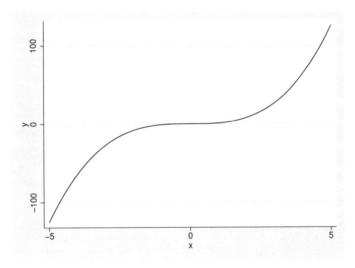

Figure 1. The function $x^3$ for $-5 \leq x \leq 5$

Otherwise put, for any $a \geq 0$, we can write

$$(-a)(-a)(-a) = -a^3 \qquad\qquad (a)(a)(a) = a^3$$

so that cube roots are defined for negative, zero, and positive cubes alike.

This concept might well be described as elementary mathematics. The volume of literature references, say, for your colleagues or students, could be multiplied; I will single out Gullberg (1997) as friendly but serious and Axler (2009) as serious but friendly. Elementary or not, Stata variously does and does not seem to know about cube roots:

```
. di 8^(1/3)
2
. di -8^(1/3)
-2
. di(-8)^(1/3)
.
. set obs 1
obs was 0, now 1
. gen minus8 = -8
. gen curtminus8 = minus8^(1/3)
(1 missing value generated)
```

This tip covers a bundle of related questions: What is going on here? In particular, why does Stata return missing when there is a perfectly well-defined result? How do we get the correct answer for negative cubes? Why should we ever want to do that? Even if you never do in fact want to do that, the examples above raise details of how Stata works that you should want to understand.

Those who know about complex analysis should note that we confine ourselves to real numbers throughout.

## 2    Calculation of cube roots

To Stata, cube roots are not special. As is standard with mathematical and statistical software, there is a dedicated square-root function sqrt(); but cube roots are just powers, and so they are obtained by using the ^ operator. I always write the power for cube roots as $(1/3)$, which ensures reproducibility of results and ensures that Stata does the best it can to yield an accurate answer. Experimenting with 8 raised to the powers 0.33, 0.333, 0.3333, and so forth will show that you would incur detectable error even with what you might think are excellent approximations. The parentheses around $1/3$ are necessary to ensure that the division occurs first, before its result is used as a power.

What you understand from your study of mathematics is not necessarily knowledge shared by Stata. The examples of cube rooting $-8$ and 8 happen to have simple integer solutions $-2$ and 2, but even any appearance that Stata can work this out as you would is an illusion:

```
. di 8^(1/3)
2
. di %21x 8^(1/3)
+1.fffffffffffffX+000
. di %21x 2
+1.0000000000000X+001
```

Showing here results in hexadecimal format (Cox 2006) reveals what might be suspected. No part of Stata recognizes that the answer should be an integer. The problem is being treated as one in real (not integer) arithmetic, and the appearance that 2 is the solution is a pleasant side effect of Stata's default numeric display format. Stata's answer is in fact a smidgen less than 2. What is happening underneath? I raised this question with William Gould of StataCorp and draw on his helpful comments here. He expands further on the matter within the Stata blog; see "How Stata calculates powers" at http://blog.stata.com/2011/01/20/how-stata-calculates-powers/.

The main idea here is that Stata is using logarithms to do the powering. This explains why no answer is forthcoming for negative arguments in the `generate` statement: because the logarithm is not defined for such arguments, the calculation fails at the first step and is fated to yield missings.

We still have to explain why `di -8^(1/3)` yields the correct result for the cube root of $-8$. That is just our good fortune together with convenient formatting. Stata's precedence rules ensure that the negation is carried out last, so this request is equivalent to `-(8^(1/3))`, a calculation that happens to have the same answer. We would not always be so lucky: for the same reason, `di -2^2` returns $-4$, not 4 as some might expect.

The matter is more vexed yet. Not only does 1/3 have no exact decimal representation, but also, more crucially, it has no exact binary representation. It is easy enough to trap 0 as a special case so that Stata does not fail through trying to calculate ln 0. But Stata cannot be expected to recognize 1/3 as its own true self. The same goes for other odd integer roots (powers of 1/5, 1/7, and so forth) for which the problem also appears.

To get the right result, human intervention is required to spell out what you want. There are various work-arounds. Given a variable y, the cube root is in Stata

```
cond(y < 0, -((-y)^(1/3)), y^(1/3))
```

or

```
sign(y) * abs(y)^(1/3)
```

The `cond()` function yields one of two results, depending on whether its first argument is nonzero (true) or zero (false). See Kantor and Cox (2005) for a tutorial if desired. The `sign()` function returns $-1$, 0, or 1 depending on the sign of its argument. The `abs()` function returns the absolute value or positive square root. The second of the two solutions just given is less prone to silly, small errors and extends more easily to Mata. The code in Mata for a scalar y is identical. For a matrix or vector, we need the generalization with elementwise operators,

```
sign(y) :* abs(y):^(1/3)
```

# 3   Applications of cube roots

Cube roots do not have anything like the utility, indeed the natural roles, of logarithms or square roots in data analysis, but they do have occasional uses. I will single out three reasons why.

First, the cube root of a volume is a length, so if the problem concerns volumes, dimensional analysis immediately suggests cube roots as a simplifying transformation. A case study appears in Cox (2004).

Second, the cube root is also a good transformation yielding approximately normal distributions from gamma or gamma-like distributions. See, for example, McCullagh and Nelder (1989, 288–289). Figure 2 puts normal probability plots for some raw and transformed chi-squared quantiles for 4 degrees of freedom side by side.

```
. set obs 99
. gen chisq4 = invchi2(4, _n/100)
. qnorm chisq4, name(g1)
. gen curt_chisq4 = chisq4^(1/3)
. qnorm curt_chisq4, name(g2)
. graph combine g1 g2
```

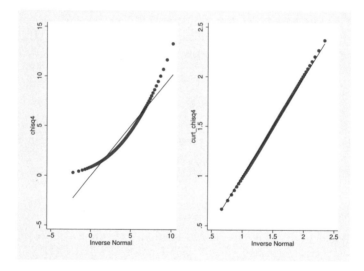

Figure 2. Ninety-nine quantiles from a chi-squared distribution with 4 degrees of freedom are distinctly nonnormal, but their cube roots are very nearly normal

The cube root does an excellent job with a distinctly nonnormal distribution. It has often been applied to precipitation data, which are characteristically right-skewed and sometimes include zeros (Cox 1992).

Third, beyond these specific uses, the cube root deserves wider attention as the simplest transformation that changes distribution shape but is easily applicable to values with varying signs. It is an odd function—an odd function $f$ is one for which

$f(-x) = -f(x)$—but in fact, it treats data evenhandedly by preserving the sign (and in particular, mapping zeros to zeros).

There are many situations in which response variables in particular can be both positive and negative. This is common whenever the response is a balance, change, difference, or derivative. Although such variables are often skew, the most awkward property that may invite transformation is usually heavy (long or fat) tails, high kurtosis in one terminology. Zero usually has a strong substantive meaning, so that we should wish to preserve the distinction between negative, zero, and positive values. (Celsius or Fahrenheit temperatures do not really qualify here, because their zero points are statistically arbitrary, for all the importance of whether water melts or freezes.)

From a different but related point of view, there are frequent discussions in statistical (and Stata) circles of what to do when on other grounds working on logarithmic scales is indicated, but the data contain zeros (or worse, negative values). There is no obvious solution. Working with logarithms of $(x + 1)$ or more generally $(x + k)$, $k$ being large enough to ensure that $x + k$ is always positive, variously appeals and appalls. Even its advocates have to admit to an element of fudge. It certainly does not treat negative, zero, and positive values symmetrically.

Other solutions that do precisely that include $\text{sign}(x) \ln(|x| + 1)$ and $\text{asinh}(x)$, although such functions may appear too complicated or esoteric for presentation to some intended audiences. As emphasized earlier, the cube root is another and simpler possibility. It seems unusual in statistical contexts to see its special properties given any mention, but see Ratkowsky (1990, 125) for one oblique exception.

One possible application of cube roots is whenever we wish to plot residuals but also to pull in the tails of large positive and negative residuals compared with the middle of the distribution around zero. See Cox (2008) for pertinent technique. In this and other graphical contexts, the main question is not whether cube roots yield approximately normal distributions, but simply whether they make data easier to visualize and to think about.

These issues extend beyond cube roots. As hinted already, higher odd integer roots (fifth, seventh, and so forth) have essentially similar properties although they seem to arise far less frequently in data analysis. Regardless of that, it is straightforward to define powering of negative and positive numbers alike so long as we treat different signs separately, most conveniently by using `sign()` and `abs()` together with the power operator. That is, the function may be defined as $-(-y)^p$ if $y < 0$, and $y^p$ otherwise.

# References

Axler, S. 2009. *Precalculus: A Prelude to Calculus*. Hoboken, NJ: Wiley.

Cox, N. J. 1992. Precipitation statistics for geomorphologists: Variations on a theme by Frank Ahnert. *Catena* 23 (Suppl.): 189–212.

———. 2004. Speaking Stata: Graphing model diagnostics. *Stata Journal* 4: 449–475.

————. 2006. Stata tip 33: Sweet sixteen: Hexadecimal formats and precision problems. *Stata Journal* 6: 282–283.

————. 2008. Stata tip 59: Plotting on any transformed scale. *Stata Journal* 8: 142–145.

Gullberg, J. 1997. *Mathematics: From the Birth of Numbers.* New York: W. W. Norton.

Kantor, D., and N. J. Cox. 2005. Depending on conditions: A tutorial on the cond() function. *Stata Journal* 5: 413–420.

McCullagh, P., and J. A. Nelder. 1989. *Generalized Linear Models.* 2nd ed. London: Chapman & Hall/CRC.

Ratkowsky, D. A. 1990. *Handbook of Nonlinear Regression Models.* New York: Dekker.

The Stata Journal (2011)
**11**, Number 2, pp. 315–317

# Stata tip 97: Getting at $\rho$'s and $\sigma$'s[1]

Maarten L. Buis
Department of Sociology
Tübingen University
Tübingen, Germany
maarten.buis@uni-tuebingen.de

There are several models in Stata that estimate coefficients other than regular "regression-like" coefficients. Often these have an interpretation as a standard deviation or a correlation of either error terms or random coefficients. Examples are `mixed` or `heckman` (see [ME] **mixed** or [R] **heckman**). Sometimes, we want access to these coefficients to perform a test or to impose a constraint. However, getting access to these parameters is not always straightforward. The problem is that Stata programs often do not estimate these coefficients directly, but instead estimate a transformed version of those parameters. In this Stata tip, I will illustrate how to recover these parameters, test hypotheses, and impose constraints.

The reason for estimating the transformed parameters rather than estimating the parameters directly is that not all positive and negative numbers represent valid values for correlations or standard deviations, which can make it more difficult to fit the model. The transformation is a way to work around this problem because this transformation is chosen such that all numbers represent valid values for this parameter. For example, a standard deviation can only be larger than or equal to 0; if we model the logarithm of the standard deviation instead, then all positive and negative numbers represent valid values. Similarly, correlations are often transformed using a Fisher's $z$ transformation (for a discussion of this transformation, see Cox [2008a]). This transformation is represented by $z = 1/2\{\ln(1 + \rho) - \ln(1 - \rho)\}$. In Stata, you would compute this as $z = $ `atanh(`$\rho$`)`. The inverses of these transformations are implemented in Stata as the `exp()` and `tanh()` functions (see [D] **functions**).[2]

The example below will use the `heckman` command because `heckman` will return both standard deviations and correlations on a transformed scale. Other commands where these strategies can be applied are, for example, `heckprobit`, `etregress`, `intreg`, `mixed`, `xtlogit`,[3] `xtprobit`,[3] `meqrlogit`, and `meqrpoisson`. Moreover, there are some programs that estimate additional parameters on a transformed scale that are not standard deviations or correlations but to which this same strategy can also be applied, for example, `nbreg` or `streg`.

If we want to recover the values of the parameters, we need to know how Stata is calling them. To find that out, we can add the `coeflegend` option to our estimation command. We can then see that neither `rho` nor `sigma` has a legend attached to it.

---

1. This tip was updated to reflect the improved multilevel mixed-effects commands introduced in Stata 13.—Ed.
2. The Mata equivalents of these functions have the same names and are documented in [M-5] **exp( )** (for `exp()` and `ln()`) and [M-5] **sin( )** (for `atanh()` and `tanh()`).
3. However, the `rho` in the output of these commands does not refer to a correlation.

However, there are parameters called `athrho` and `lnsigma` that do have legends attached
to them, and we can read in the manual (and deduce from their names) that these are the
Fisher's $z$ transformed correlation and the natural logarithm of the standard deviation,
respectively. So we can recover these parameters as follows:

```
. use http://www.stata-press.com/data/r11/womenwk

. heckman wage educ, select(married children educ) nolog coeflegend
```

```
Heckman selection model                          Number of obs    =       2000
(regression model with sample selection)         Censored obs     =        657
                                                 Uncensored obs   =       1343

                                                 Wald chi2(1)     =     403.39
Log likelihood = -5250.348                       Prob > chi2      =     0.0000
```

| wage | Coef. | Legend |
|---:|---:|---|
| **wage** | | |
| education | 1.099506 | _b[wage:education] |
| _cons | 7.042147 | _b[wage:_cons] |
| **select** | | |
| married | .5420304 | _b[select:married] |
| children | .4409418 | _b[select:children] |
| education | .0722993 | _b[select:education] |
| _cons | -1.473038 | _b[select:_cons] |
| /athrho | .8081049 | _b[athrho:_cons] |
| /lnsigma | 1.807547 | _b[lnsigma:_cons] |
| rho | .6685435 | |
| sigma | 6.095479 | |
| lambda | 4.075093 | |

```
LR test of indep. eqns. (rho = 0):   chi2(1) =     47.02   Prob > chi2 = 0.0000
```

```
. // standard deviation of the residual of the wage equation
. di exp([lnsigma]_b[_cons])
6.0954785

. //correlation between residuals of the wage and selection equation
. di tanh([athrho]_b[_cons])
.6685435
```

If we have specific hypotheses, then one way of testing these is to rephrase these hypotheses in terms of the transformed metric. Here I compute the transformed standard
deviations and correlations on the fly by using the trick to enter the computations as
'= *exp*' (Cox 2008b). This way, *exp* is immediately evaluated, and `test` only sees the
number that is the result of that computation. Notice that the transformations are not
defined for standard deviations of 0 or correlations of $-1$ or 1. This is another way in
which we can see that one needs to be careful when testing hypotheses on the boundary
of the parameter space (for example, see Gutierrez, Carter, and Drukker [2001]).

```
. test ([lnsigma]_b[_cons] = `= ln(6)´    )
>      ([athrho]_b[_cons]  = `= atanh(.7)´)
 ( 1)  [lnsigma]_cons = 1.791759
 ( 2)  [athrho]_cons = .8673005
           chi2(  2) =    2.29
         Prob > chi2 =    0.3177
```

Using similar tricks, we can also impose constraints on these transformed auxiliary parameters.

```
. constraint 1 [lnsigma]_cons = `= ln(6)´
. constraint 2 [athrho]_cons  = `= atanh(.7)´
. heckman wage educ, select(married children educ) constraint(1 2) nolog
```

| Heckman selection model | | | | | Number of obs | = | 2000 |
| (regression model with sample selection) | | | | | Censored obs | = | 657 |
| | | | | | Uncensored obs | = | 1343 |
| | | | | | Wald chi2(1) | = | 465.26 |
| Log likelihood = -5251.522 | | | | | Prob > chi2 | = | 0.0000 |

```
 ( 1)  [lnsigma]_cons = 1.791759
 ( 2)  [athrho]_cons = .8673005
```

| wage | Coef. | Std. Err. | z | P>\|z\| | [95% Conf. Interval] | |
|---|---|---|---|---|---|---|
| **wage** | | | | | | |
| education | 1.104284 | .0511957 | 21.57 | 0.000 | 1.003943 | 1.204626 |
| _cons | 6.911303 | .7037765 | 9.82 | 0.000 | 5.531927 | 8.29068 |
| **select** | | | | | | |
| married | .5405253 | .0639579 | 8.45 | 0.000 | .4151702 | .6658805 |
| children | .4405534 | .0258773 | 17.02 | 0.000 | .3898347 | .491272 |
| education | .0725339 | .010469 | 6.93 | 0.000 | .0520151 | .0930527 |
| _cons | -1.467094 | .1436783 | -10.21 | 0.000 | -1.748699 | -1.18549 |
| /athrho | .8673005 | . | . | . | . | . |
| /lnsigma | 1.791759 | . | . | . | . | . |
| rho | .7 | . | | | -1 | 1 |
| sigma | 6 | . | | | . | . |
| lambda | 4.2 | . | | | . | . |

```
LR test of indep. eqns. (rho = 0):   chi2(1) =    44.67   Prob > chi2 = 0.0000
```

# References

Cox, N. J. 2008a. Speaking Stata: Correlation with confidence, or Fisher's z revisited. *Stata Journal* 8: 413–439.

———. 2008b. Stata tip 59: Plotting on any transformed scale. *Stata Journal* 8: 142–145.

Gutierrez, R. G., S. Carter, and D. M. Drukker. 2001. sg160: On boundary-value likelihood-ratio tests. *Stata Technical Bulletin* 60: 15–18. Reprinted in *Stata Technical Bulletin Reprints*, vol. 10, pp. 269–273. College Station, TX: Stata Press.

# Stata tip 98: Counting substrings within strings

Nicholas J. Cox
Department of Geography
Durham University
Durham, UK
n.j.cox@durham.ac.uk

Consider the following problem, based on a real one reported on Statalist. A user has string date–times in differing forms; let us imagine they are in a variable, sdate. The desire is for everything to be DMY hms—that is, day, month, year, space, hours, minutes, and seconds. Examples of data are

> "25/12/2010 11:22"
> "25/12/2010 11:22:33"
> "25/12/2010 11:22:33:444"

Compared with the standard, the first example lacks seconds; the second example is fine; and the third example includes milliseconds, but expressed in a nonstandard way. Stata expects seconds and milliseconds in the decimal form "33.444".

Some cleanup is required. One way forward is to note that the number of colons (:) present within the string is diagnostic of whether action is required and what to do. So how do you count substrings (in this case, just a colon) within strings? The problem has been aired in this journal at least once before (Cox 2011), but it occurs sufficiently often that it deserves a further flag. The main solution tends to provoke the reaction "Yes, of course!", whether because people know it already or they now see that it is obvious and direct. However, people (including myself) have often supposed that the problem requires a more complicated approach than it really does.

Faced with this kind of problem, an experienced user tends to browse the help for string functions to see if there is a function dedicated to this problem, but in this case browsing will be in vain. But as so often happens, combining different functions is more successful. Consider how you would count for yourself. You would naturally work from one end of the string to the other, noting each occurrence. It is immaterial whether you count from left to right or from right to left, but note that Stata, by default, works on strings from left to right.

Stata has a function, subinstr(), that looks for occurrences of substrings within strings and replaces them with a specified substring (often just an empty string, ""). This function gives us a solution. Consider the calculation

```
length(sdate) - length(subinstr(sdate, ":", "", .))
```

As with elementary algebra, working from the inside of complicated expressions outward is a good tactic for understanding. The function call

```
subinstr(sdate, ":", "", .)
```

takes in this example a named variable, `sdate`, and substitutes an empty string, `""`, for every occurrence of the single-character substring that is a colon, `":"`. In other words, it deletes every colon. The length of the result is shorter than the original by how many colons were found, which could be 0, 1, 2, 3, and so forth.

We do not always need to get Stata to do this and remove those substrings. In this example, and in many others, doing so would just mess up our data and make our problem more difficult. We just need to get Stata to tell us what the result would be. So, to conclude the example, we could count the colons, fix problem cases, and check to see if that strategy worked everywhere:

```
generate ncolons = length(sdate) - length(subinstr(sdate, ":", "", .))
replace sdate = sdate + ":30" if ncolons == 1
replace sdate = reverse(subinstr(reverse(sdate), ":", ".", 1)) if ncolons == 3
generate double ndate = clock(sdate, "DMY hms")
format ndate %tC
list sdate if missing(ndate)
```

The trickiest detail here is that to change the last colon of three to a period, we reverse the string, change the first colon, and reverse it back again. This is another common example of combining string functions.

Counting substrings that are two or more characters long introduces two extra twists to the problem.

First, let us focus on a standard elementary puzzle. How many occurrences of `"ana"` are there in `"banana"`? One tricksy answer is two—namely, `"*ana**"` and `"***ana"`—but Stata's answer using the `subinstr()` trick will have to be one. Once the first `"ana"` is blanked out, `"bna"` is all that is left. It seems rare in data management that we want the more generous answer, but how would we do it?

In essence, we need to test each possible position. For that, we need to know the length of the string—in our example, a string variable (say, `svar`)—to be searched. Either we know the precise type of a string variable in advance—say, `str12`—or we will need to look it up using an extended macro function. Let's choose the second and marginally more difficult case for a string variable. We will keep going with the silly example of counting as many occurrences of `"ana"` as possible.

```
local length = substr("`: type svar'", 4, .) - length("ana") + 1
gen ana_count = 0
qui forval i = 1/`length' {
        replace ana_count + (strpos(svar, "ana") == `i')
}
```

How does this approach work? We are testing each possible position. We start at the first character, but we need not go all the way up to the end, because for example, the last possible position in which `"ana"` could occur within a `str12` is position 10 (not position 9!) We initialize a count variable as 0 and bump it up by 1 every time the position is indeed a starting position for the substring. That follows from the expression (`strpos(svar, "ana")` == `i`) evaluating to 1 when true and 0 when false, so we end up counting occurrences.

That case is clearly more complicated than the case of counting occurrences disjointly. Fortunately, it appears to arise much less frequently.

Second, remember to adjust for the length of substring when counting substrings using the `subinstr()` method. `length("banana")` - `length(subinstr("banana",` `"ana", "", .))` yields 3 because three characters were removed, but just one occurrence of the substring was. In general, divide by `length("`*substring*`")`. If you know what it is, there is a small efficiency gain in saying so. For example, you should not write

```
gen n_ana = (length(svar) - length(subinstr(svar, "ana", "", .))) / length("ana")
```

because that obliges Stata to calculate `length("ana")` for every observation. Even in a programming situation where the particular substring is not predictable in advance, a calculation like

```
local sslen = length("substring")
```

allows the result to be used in macro form in a command:

```
gen n_substring = (length(svar) - ///
   length(subinstr(svar, "substring", "", .))) / `sslen´
```

That way, the contents of the macro are substituted before the **generate** command gets to work so that it uses a known constant rather than an expression to be evaluated for every observation.

The problem could get more complicated, yet. For example, the substring concerned might vary from observation to observation, as when conventions about reporting are variable in the data. So long as the substring is known and included within a variable (say, `ssvar`), this is no more difficult than any previous problem.

```
gen sscount = (length(svar) - ///
   length(subinstr(svar, ssvar, "", .))) / length(ssvar)
```

Alternatively, we might be counting different possible types of substring, for which we just need to cycle over all the possibilities.

# Reference

Cox, N. J. 2011. Speaking Stata: MMXI and all that: Handling Roman numerals within Stata. *Stata Journal* 11: 126–142.

The Stata Journal (2011)
**11**, Number 2, pp. 321–322

# Stata tip 99: Taking extra care with encode

Clyde Schechter
Albert Einstein College of Medicine
Yeshiva University
New York, NY
clyde.schechter@einstein.yu.edu

encode (see [D] **encode**) has long been one of Stata's basic data-management commands. encode maps the distinct strings of a string variable to an integer-valued numeric variable for which the strings become value labels. Unless you specify a preexisting set of value labels through its label() option, encode uses the alphanumeric order of distinct string values present in the dataset to determine numeric values 1, 2, 3, and so on. Thus if "a", "b", and "d" were the distinct values of a variable, svar, in one dataset, then typing

```
. encode svar, generate(nvar)
```

would produce **nvar**, in which 1, 2, and 3 correspond to "a", "b", and "d", respectively. However, the same command applied to a dataset in which "a", "b", "c", and "d" were the distinct values of **svar** would produce an encoding that was overlapping but also different: 3 would correspond to "c" and 4 to "d". Because value labels are what the user sees in text and graphic output, it could be easy to miss the difference on casual inspection. Moreover, this difference could easily prove problematic if two or more datasets were to be combined, say, by using append or merge. Indeed, using encode on the same variable in multiple datasets that will later be combined can only be called dangerous.

Having been bitten by this many times, I have developed some precautionary data-management practices that I commend to others.

1. There are certain types of variables that recur frequently in my work. For many of these variables, I have developed a standard encoding that I always use. The code to create standard value labels is explicit in some do-files that I routinely do, run, or include in my dataset creation do-files. These value labels cover all the possible values these variables can take. Whenever I encode one of these variables, I always explicitly use the label() option with these labels.

2. In large projects that will involve multiple datasets with overlapping variables not part of my "standard" list, whenever I use encode, I routinely follow up with a label save as an audit of that particular encoding. In later work with the same variable in other datasets, before I encode, I again do, run, or include the corresponding labeling do-file and then use the explicit label() option in the encode command. If encode finds new levels of the variable not already in the label, it adds them to the label. I follow up using label save, replace again so that my labeler do-file remains up to date.

3. So that I do not rely on my memory to know whether I have previously developed a labeling for a variable, my practice for nonroutine variables is to give the value label the same name as the variable and to name the labeler do-file using the form *varname*_label.do. Then, when I want to encode such a variable, I precede the encode with

```
. capture run varname_label.do
```

In fact, I have an ado-file that is a wrapper for encode—it handles all this for me.

Although these practices may seem cumbersome and can lead to a project directory being a bit cluttered with do-files that just generate labels, adherence to these practices has saved me from some nasty analysis errors that are hard to root out otherwise.

The Stata Journal (2011)
11, Number 2, pp. 323–324

# Stata tip 100: Mata and the case of the missing macros

William Gould
StataCorp
College Station, TX
wgould@stata.com

Nicholas J. Cox
Department of Geography
Durham University
Durham City, UK
n.j.cox@durham.ac.uk

People who come from Stata programming to Mata often look around and ask, "Where is the macro substitution?" The short and discouraging answer is that there is none. The longer and much more encouraging answer is that what you would do by macro substitution in Stata is easily done in other ways in Mata.

This tip presupposes some acquaintance with macros in Stata. If macros are new or not very familiar to you, [U] **18 Programming Stata** will give you a lot of background.

Let's make clear what we mean with an example. Consider a loop in Stata that runs

```
forvalues j = 1/42 {
        summarize P`j´
}
```

forvalues (see [P] **forvalues**) loops over integer sequences, here all of them from 1 to 42. So as Stata executes the loop, the local macro j takes on the values 1, 2, ..., 41, 42, which are substituted in turn within the text P'j'. The effect will be that Stata sees, in turn,

```
summarize P1
summarize P2
```

and so on, up to summarize P42. Assuming that you do have variables P1–P42, then each will be summarized in turn.

Now let us suppose that we want to do something similar in Mata. As an artificial example sufficient to show the main idea, let's read in those variables and spit out their means from Mata. However, Mata has no idea of a local or global macro. If it sometimes appears to understand such macros, that is only because references to them are interpreted by Stata before they are ever seen by Mata. But we can do the same kind of string manipulation in Mata, using its own functions.

Here is one solution:

```
for(j = 1; j <= 42; j++) {
        name_to_use = sprintf("P%g", j)
        mean(st_data(., name_to_use))
}
```

The integer j is mapped to its string equivalent by being inserted as one or more characters in a string that could be displayed. We are not going to display that string, but we are going to use its contents, which in the same way will be in turn "P1", "P2", and so on.

Here is another way to get `name_to_use` within the same loop:

```
name_to_use = "P" + strofreal(j)
```

It does not matter which command line appears in code. Either way, the idea so far is to construct the desired variable name in Mata as a string scalar, `name_to_use`. In each case, look at the help (equivalently, the Mata manual) if you want more detail on functions such as `sprintf()` or `strofreal()`.

But we can short-circuit that longer code by combining two lines of code into one, which incidentally gets us even closer to the idea of string substitution:

```
mean(st_data(., sprintf("P%g", j)))
```

or

```
mean(st_data(., "P" + strtoreal(j)))
```

The difference is just a matter of style. Some users will prefer the step-by-step solution. The choice is yours. So the main idea is simple: use Mata's display or string functions to do the manipulations you need.

The Stata Journal (2011)
11, Number 3, pp. 472–473

# Stata tip 101: Previous but different

Nicholas J. Cox
Department of Geography
Durham University
Durham City, UK
n.j.cox@durham.ac.uk

Given a time series that changes intermittently, there may be interest in identifying the most recent value that differed from the present value. Given a history that was 36, 36, 36, 42 before a present value of 42, the value that was most recent but different is 36. That is easy enough to determine by eye, but how could we calculate that systematically in Stata? Such distinct values define runs or spells, as discussed in detail in Cox (2007), but here we examine the problem directly.

Adding some more detail, let us suppose first that we have a single panel with data starting with

```
time:  2, 3, 5, 7, 11
value: 36, 36, 36, 42, 42
```

The example is of irregularly spaced times, and in fact only one time step is 1 in whatever units we are using. These data were devised to emphasize that the method to be explained does not depend on anything other than the time variable being known and single-valued for a given panel and values being in `sort` order of time. We can always ensure that last condition by `sorting` on the time variable. Not only do we not need to `tsset`, but that machinery and the related apparatus of time-series operators (such as `L.` and `D.`) would not help much, even for regularly spaced data.

Information on previous values comes from when the value changes. We can create a new variable containing previous values using subscripts:

```
. generate previous = value[_n-1] if value != value[_n-1]
```

Because `_n` means the present observation number, $\_n - 1$ means the previous observation number. If this is new to you, get more information by typing `help subscripting` into Stata.

The new variable in our example now includes missings in all but one observation: `.`, `.`, `.`, `36`, `.` are its values. The value of 36 clearly comes from the one point at which `value` changes, but we also need to understand the logic for the other observations.

Let us focus first on `value[1]`. Its previous value is `value[0]`, a hypothetical value beyond the dataset. Stata does not know what that is but does not regard a request to access `value[0]` as illegal; it merely returns missing. Now we have `value[1]` of 36, which differs from `value[0]` of missing, so `previous[1]` is returned as `value[0]`, namely, missing.

The reasoning for `value[2]`, `value[3]`, and `value[5]` is different. In each case, `value` is equal to the previous `value`, but the `generate` statement included no instructions for what to do if the inequality was not satisfied. So missing is also returned, the same result for a different reason.

We must now fill in missing values to the best of our ability. In particular, 36 is the previous but different value for observation 5 as well as observation 4. Exploiting the fact that `generate` and `replace` use the current sort order (Newson 2004), we can fill in missing values in a cascade:

```
. replace previous = previous[_n-1] if missing(previous)
```

That still leaves a block of missings before the first value change. Furthermore, if `value` never changed within our data, there would be no previous but different value to identify.

```
. list
```

|  | time | value | previous |
|----|------|-------|----------|
| 1. | 2 | 36 | . |
| 2. | 3 | 36 | . |
| 3. | 5 | 36 | . |
| 4. | 7 | 42 | 36 |
| 5. | 11 | 42 | 36 |

Now we need to extend this to any number of panels. Given also an identifier, the extension is just to do all using `by:` (Cox 2002):

```
. by id (time), sort: generate previous = value[_n-1] if value != value[_n-1]
. by id: replace previous = previous[_n-1] if missing(previous)
```

If there are missings in `value`, they would be better filled in first in a clone of `value` using the cascade device. The difference from the previous different value is then simply (regardless of panel context or time spacing)

```
. generate diff = value - previous
```

If we wanted to look forward in time to identify the next value that is different, it is easiest to reverse time temporarily:

```
. gsort id -time
. by id: generate next = value[_n-1] if value != value[_n-1]
. by id: replace next = next[_n-1] if missing(next)
. sort id time
```

# References

Cox, N. J. 2002. Speaking Stata: How to move step by: step. *Stata Journal* 2: 86–102.

———. 2007. Speaking Stata: Identifying spells. *Stata Journal* 7: 249–265.

Newson, R. 2004. Stata tip 13: generate and replace use the current sort order. *Stata Journal* 4: 484–485.

The Stata Journal (2011)
**11**, Number 3, pp. 474–477

# Stata tip 102: Highlighting specific bars

Nicholas J. Cox
Department of Geography
Durham University
Durham City, UK
n.j.cox@durham.ac.uk

A frequent need when drawing a bar or dot chart is to highlight a subset of observations while keeping the overall sort order. The stipulation of keeping the overall sort order is what provides the challenge here, because otherwise we could just add subdivision by another variable to the command, as when distinguishing foreign cars among those with the best repair record:

```
. sysuse auto
. graph hbar (asis) mpg if rep78 == 5, over(make, sort(1) descending)
. graph hbar (asis) mpg if rep78 == 5, over(make, sort(1) descending)
> over(foreign) nofill
```

Figure 1 shows the graphs for these two commands.

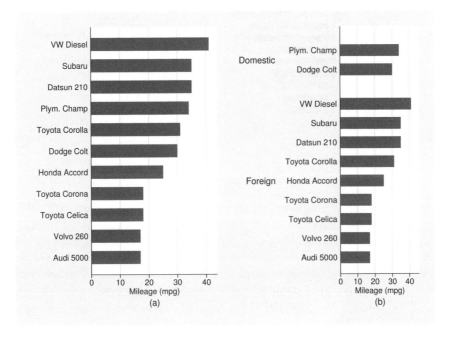

Figure 1

In figure 1(a), the ordering is within all the observations specified. In figure 1(b), the extra option `over(foreign)` subdivides observations according to the further variable `foreign`. Note also the crucial detail of `nofill`. This can be a useful kind of graph, but it is not what we want here.

Let us suppose we have data on basin (catchment or watershed) areas for various large rivers in the world, and we want to show where the Mississippi falls in the rank order for the very largest basins. Some example data from Allen (1997) are included with the media for this issue.

```
. use rivers
```

Figure 2 as a first graph shows that the Mississippi ranks third on area of basin in this dataset, after the Amazon and Zaire.

```
. graph hbar (asis) area if area >= 1000, over(name, sort(1) descending)
```

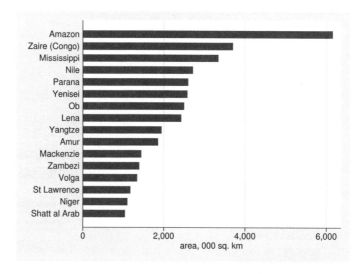

Figure 2

Highlighting a particular bar means giving it a different color. Some acquaintance with the bar chart commands shows that they are willing to combine bars for different variables, which will be assigned different colors. Knowing this, we simply need to put data for two subsets, the Mississippi and the others, into two different variables. `separate` (see [D] **separate**) is a command designed for precisely this purpose. For other graphical applications of `separate`, see Cox (2005). It is naturally also possible to use `generate` directly.

```
. separate area, by(name == "Mississippi")
```

In this example, the equality supplied to `by()` is either false or true, numerically 0 or 1, and so `separate` creates two new variables, `area0` and `area1`.

```
. graph hbar (asis) area0 area1 if area >= 1000, nofill
> over(name, sort(area) descending) legend(off) ytitle("`: var label area´")
```

We are plotting bars for values that are nonmissing on `area0` and missing on `area1`, or vice versa. But `graph` plots no bars when values are missing. This is easy to fix: `nofill` gets us the intended effect. In this case, we suppressed the legend, imagining that, depending on the purpose, we could add a title for a presentation—as, say

```
title(Mississippi ranks third in catchment area)
```

or underline the message of the graph in informative text supplied in a text or word processor. Because two response variables are being shown on the same graph, we have to step in to provide an informative *y*-axis title, in this case by automating use of the variable label for `area`. Nothing stops us from just providing an axis title explicitly, as when no such variable label has been defined.

In principle, using `stack` should have the same effect as using `nofill`. In practice, small complications can exist if there are other missing values in the data; these complications are fixable with an appropriate `if` exclusion.

The main problem now being solved, we could clearly heighten the contrast by adding `bar(1, bfcolor(none))`. Figure 3 shows the graph after that tweak.

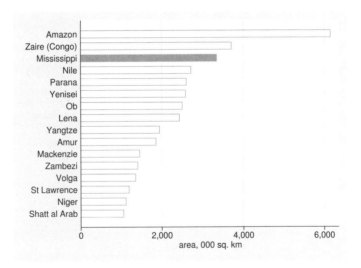

Figure 3

Similar needs are met by variations on this theme.

In our example, the subset to be highlighted is a single observation, but nothing depends on that being true.

Equally, three or more subsets could be distinguished. For a more elaborate subdivision, we might want a legend, although there is a trade-off: the more complicated the design, for which a legend becomes necessary, the less the impact of the graph is likely to be.

The examples all are based on showing values `asis`. If graphs of this kind are needed, but for means or other summary statistics, it is often easiest to `collapse` or `contract` the dataset first and then use `separate` and `graph hbar` (`asis`).

The same device can be used with `graph bar`, `graph dot`, or various subcommands of `twoway`, such as `twoway bar`. In practice, when we want this, the individual observations include names that are informative, so horizontal alignment makes those names more readable. If `graph dot` were to be used, we should consider heightening the contrast by adding, for example, `marker(2, msize(*3))`.

# References

Allen, P. A. 1997. *Earth Surface Processes*. Oxford: Blackwell Science.

Cox, N. J. 2005. Stata tip 27: Classifying data points on scatter plots. *Stata Journal* 5: 604–606.

The Stata Journal (2011)
**11**, Number 4, pp. 627–631

# Stata tip 103: Expressing confidence with gradations

Ulrich Kohler
Wissenschaftszentrum Berlin (WZB)
Berlin, Germany
kohler@wzb.eu

Stephanie Eckman
Institute for Employment Research
Nuremberg, Germany
stephanie.eckman@iab.de

Researchers often express the uncertainty associated with a parameter estimate $\widehat{\mu}$ by plotting the 95% confidence interval (CI) around the statistic. The meaning of these intervals is that in the long run, 95% of the intervals formed in this way will include the fixed but unknown parameter of interest $\mu$. It is also true, though we often neglect to mention it, that "the unknown $\mu$ is more often captured near the center of an interval than near the lower or upper limit, or end-point, of an interval" (Cumming 2007, 90). Our technique provides a method of plotting CIs that makes this facet of their interpretation more clear. Below we describe how to use Stata to plot the entire CI function—the $p$-value function (Poole 1987)—with gradations, thus depicting this often forgotten fact about CIs.

Let us start by reconsidering the creation of a plot of a point estimate with its 95% CI. The first step is usually to create a dataset containing several point estimates and their respective standard errors (what Newson [2003] calls a resultsset). We demonstrate how to do this for the arithmetic mean of the body mass index (BMI) from the National Health and Nutrition Examination Study (NHANES):

```
. use http://www.stata-press.com/data/r13/nhanes2
. collapse (mean) mean=bmi (semean) se=bmi, by(female)
```

The upper and lower bounds of the 95% CIs are calculated by adding and subtracting 1.96 times the standard error to the point estimates. We generate variables holding those values,

```
. generate upper = mean + 1.96*se
. generate lower = mean - 1.96*se
```

and use a range plot overlaid with a scatterplot to show the upper and lower limits of the CI along with the point estimate (see figure 1). Such graphs are commonly used to convey uncertainty in point estimates.

```
. graph twoway
>    || rcap upper lower female, lstyle(p1)
>    || scatter mean female, mstyle(p1)
>    || , legend(off) xlabel(0 "Men" 1 "Women") xtitle("")
>    xscale(range(-.5 1.5)) ytitle("BMI")
```

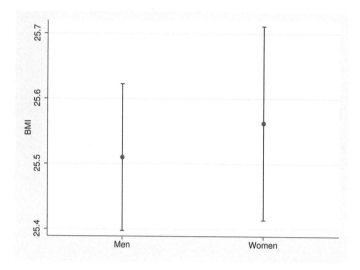

Figure 1. Average BMI of men and women with 95% CIs

The CIs shown in this graph are correct, assuming that the standard errors were calculated appropriately. However, we worry, in the spirit of Cumming (2007) and Läärä (2009, 141), that the figure gives the incorrect impression that the true value is equally likely to lie at any point in the interval. To avoid this misinterpretation, our technique plots the CIs with shading gradations to convey the sense that the true value is much more likely to be near the estimated mean than at the ends of the intervals. Our revised figure is shown in figure 2. Below we discuss the details of how we created these plots.

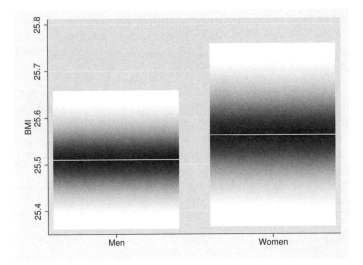

Figure 2. Average BMI of men and women with 95% CIs; the relative chance of capturing the true average is approximated by shading

In calculating the upper and lower limits of the 95% CI above, we multiplied the standard error by 1.96, the critical value of a Student's $t$ distribution for 95% CI. We can calculate this critical value in Stata by using the `invttail()` function with the appropriate degrees of freedom:

```
. display abs(invttail(10350,(1-95/100)/2))
1.9601932
```

If we insert values other than 95 into this command, we can find the critical values for constructing intervals for any given level of confidence $C$. In fact, we can use a loop to form a series of pairs of variables holding the boundaries for CIs for all values of $C = 1, 2, \ldots, 99$:

```
. forv C = 1(1)99 {
  2.     gen ub`C' = mean + abs(invttail(10350,(1-`C'/100)/2))*se
  3.     gen lb`C' = mean - abs(invttail(10350,(1-`C'/100)/2))*se
  4. }
```

We can now plot each of these 99 pairs of CI bounds in one plot. We start with the 99% CI, which is the widest one. We then overlay it with the second widest, the 98% CI, and so on. To create the gradations in figure 2, we plot the wide intervals in a lighter shade than the narrower ones.

To accomplish these shading gradations, we use plot type `rbar` or `rarea`, which allows us to change the outlook of the bars or areas with each CI we plot, using the `fcolor()`, `fintensity()`, and `lcolor()` options. The `fcolor()` option sets the fill color of the bars; for example, `fcolor(black)` makes the bars black. The `fintensity()` option sets the intensity of the selected colors with a parameter from 0 to 100; for example, `fintensity(0)` sets a low intensity and `fintensity(100)` sets a high intensity. The `lcolor()` option sets the outline color; for example, `lcolor(black)` outlines the bars or areas with black lines. It is also possible to control the intensity of the outline color by multiplying the color with an intensity; for example, `lcolor(black*0.9)` is the same color as `fcolor(black)` with `fintensity(90)`.

Thankfully, we do not need to write the command for each of the 99 plots by hand. Instead, we can use a loop to create a local macro that holds the code for all 99 CI plots. We then create the command for the graph itself with that local macro. Here is the loop to create the local macro `rbar`, which holds the definitions for the 99 `rbar` plots. Note the use of the local `i` inside the loop to change the intensity and the line color with each plotted CI.

```
. forvalue i = 99(-1)1 {
  2.        local  rbarvar `rbarvar´
  >         || rbar ub`i´ lb`i´ female,
  >         fcolor(black) fintensity(`=100-`i´´)
  >         lcolor(black*`=(100-`i´)/100´)
  >         barwidth(.8)
  3. }
```

And here is how we use the local macro `rbarvar` in a `graph twoway` command to create the graph. (The `graph` command also uses `scatteri` to add white lines to indicate the point estimates. We set the background color of the plot region to `gs14` to make the bright parts of the plot stand out.)

```
. graph twoway
>    `rbarvar´
>    || scatteri `=mean[1]´ -.4 `=mean[1]´ +.4
>    , recast(line) lcolor(white) lpattern(solid)
>    || scatteri `=mean[2]´ .6 `=mean[2]´ +1.4
>    , recast(line) lcolor(white) lpattern(solid)
>    || , legend(off) xlabel(0 "Men" 1 "Women") xtitle("")
>    plotregion(color(gs14)) ytitle("BMI")
```

The result is figure 2, shown above. The range bars spread between the limits of the 99% CI; however, the relative chance of capturing the true mean value of the BMI is approximated by shading: the darker the color, the greater the relative chance that the true mean lies in that area. Although "the vagaries of printing, and the human visual system, mean that [the plot] may not give an accurate impression of the relative chance" (Cumming 2007, 90), we feel that figure 2 does a better job of conveying the meaning of a CI than does figure 1.

This technique generalizes to many other situations. It works well for other kinds of point estimates and can also display CIs based on standard errors calculated by jackknife, bootstrap, or other methods.

One note of caution: We advise users of this technique to work with the resultsset approach (Newson 2003), that is, to construct a dataset that holds only the point estimates and their standard errors. Otherwise, the creation of 99 pairs of CI boundaries in large datasets could cause a user to run into memory issues.

# References

Cumming, G. 2007. Inference by eye: Pictures of confidence intervals and thinking about levels of confidence. *Teaching Statistics* 29: 89–93.

Läärä, E. 2009. Statistics: Reasoning on uncertainty, and the insignificance of testing null. *Annales Zoologici Fennici* 46: 138–157.

Newson, R. 2003. Confidence intervals and p-values for delivery to the end user. *Stata Journal* 3: 245–269.

Poole, C. 1987. Beyond the confidence interval. *American Journal of Public Health* 77: 195–199.

The Stata Journal (2011)
**11**, Number 4, pp. 632–633

# Stata tip 104: Added text and title options

Nicholas J. Cox
Department of Geography
Durham University
Durham, UK
n.j.cox@durham.ac.uk

A frequent desire in graph production is to add explanatory text at appropriate places on a graph. It is a good idea in general to do that sparingly. Similarly, it may be a bad idea in particular to do that at all if a graph is already crowded and likely to prove challenging to readers.

Scatterplots and line plots are the most common examples for which such annotation is wanted. For identifying or commenting on outliers or other particular data points, it is natural to turn to marker labels or added text options, documented in [G-3] *marker_label_options* and [G-3] *added_text_options*. Both kinds of options share the property that they are linked to particular locations specified in terms of the two axes of the graph in the units of the two variables being shown. Sometimes that is exactly the control you need, but it is awkward if what you want is just to put text according to relative position within the data region. Usually, you might want to put the extra text in a corner of the graph. The difficulty then is that Stata's choices of axis limits depend not only on any axis scale options that may be set, but also on the ranges of the variables, what axis labels and ticks have been requested explicitly and implicitly, and so on. Optimizing choices for a single graph can be annoying but is likely to be tolerable. Optimizing choices for several graphs is much more awkward, especially if the aim is to automate a series of graphs though some of the variables' properties are not known exactly in advance.

A point sometimes missed, and the main reason for this tip, is that title and legend options can be moved away from their default positions and placed within the data region. The presumption here is that they are not needed for other purposes. However, there are several title options to choose from, and it is likely that at least one is not currently used. (If you are using all of the options `title()`, `subtitle()`, `note()`, and `caption()`, to say nothing of others that are available, then your graph may be too complicated already.)

The main idea is very simple. The default positions under the graph scheme are precisely that: default choices, or suggestions that may be altered. If you use one of these options, then other defaults may need to be modified too, but that is generally straightforward.

A single example should be enough to show the idea. Imagine that we want to comment generally on the merits of a fitted line.

```
. sysuse auto
(1978 Automobile Data)

. scatter mpg weight, ms(Oh) || lfit mpg weight, lc(gs12)
>      note("note that the linear fit" "misses the curvature", ring(0)
>            pos(1) size(*1.6) box)
>      legend(off) ytitle("`: var label mpg'")
```

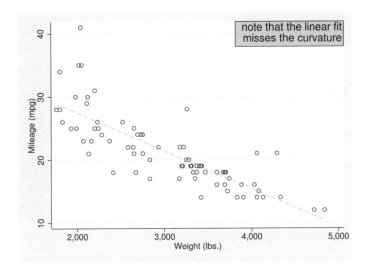

Here we used the `note()` option, varying not only where the note is placed (`ring(0)` `pos(1)`) but also its size and whether it is boxed. The example also shows that multiline comments, a common wish, are easy to produce.

With such tricks, automated production of a series of graphs in a uniform style is much easier to achieve. In this example, the overall correlation is negative, and the top right or northeast corner is a natural place for the extra text. When showing (for example) an increasing time series, the top left corner will seem natural, and other such variations will quickly spring to mind.

As mentioned above, legend options may be used too. In fact, surprising though it may seem, the legend need make no references to any of the variables being shown in the graph, as detailed study of the documentation will make clear.

This tip picks no quarrel with users who prefer to keep such extra text outside the data region or confine it to the caption that will usually be composed as part of the text. The idea is just to indicate what choices are available without moralizing too much about choices you must or should make.

The Stata Journal (2012)
**12**, Number 1, pp. 159–161

# Stata tip 105: Daily dates with missing days

Steven J. Samuels
18 Cantine's Island
Saugerties, NY 12477
USA
sjsamuels@gmail.com

Nicholas J. Cox
Department of Geography
Durham University
Durham, UK
n.j.cox@durham.ac.uk

## 1   Introduction

In projects that record daily dates, there may be some observations for which only the month and year are known. This can happen, for example, when memories of events are fuzzy or written records were never kept or have been lost.

In this tip, we suggest some simple strategies for dealing with missing daily dates. Whatever is done should, naturally, be done cautiously. In addition to solving a real problem met in data handling, the strategies we suggest here provide a good exercise in combining Stata date functions. For the sake of clarity, we work with more new variables than is strictly necessary. Toward the end of the tip, we will explain how one new variable should be kept as a record of what was done.

The process of handling complete dates is easy. If the year, month, and day are recorded as separate variables—say, `year`, `month`, and `day`—we can

```
. generate dailydate = mdy(month,day,year)
```

If the date is imported in a string variable—say, `reported`—either delimited (as in `"2008/1/13"`) or run together (as in `"20080113"`), we can use Stata's powerful date parser:

```
. generate dailydate = date(reported, "YMD")
```

These `generate` commands will return missing dates if the day is missing. We still want to use the information we have for month and year. If we have read in a string variable, we can create `year` and `month` numeric variables to hold the known year and month. For example, if the date is run together as above, then

```
. generate month = substr(reported,1,4)
. generate year = substr(reported,5,.)
```

This puts the two forms of input (separate numeric variables and strings) on equal footing.

## 2   If a monthly date is enough

In some situations, the analyst might decide that knowing the month of the event is sufficient. For example, the event might be a medical test that should be performed every month. If so, we can create a monthly date with Stata's `monthly()` function:

```
. generate int monthlydate = monthly(year + "-" + month, "YM")
```

For "200801", this recipe yields 576 as a monthly date (January 1960 is month 0). If `monthlydate` is given a `%tm` format, it will display as "2008m1".

## 3   Use a midpoint date

If months are not enough, then a common work-around is to substitute 15 for the missing day. In a nonleap year, the mean number of days in each month is

$$(4/12) \times 30 + (7/12) \times 31 + (1/12) \times 28 = 30.416667$$

In a leap year, it is 30.5. Therefore, on average, 15 is the integer closest to the mean. With this choice, we can be assured that the error in each date is no more than $\pm 16$ days in any month.

```
. generate imputedday = 15
```

Substitution of 15 for the missing day is an example of *mean imputation* (Little and Rubin 2002; Seastrom, Kaufman, and Lee 2002). In general, mean imputation is undesirable because it produces spikes at the mean and distorts distributional parameters such as the variance. This undesirable behavior can also carry over to the distribution of intervals between imputed dates. For example, the absolute difference of two dates, one imputed to be the 15th, in one month will have a range no more than 16 days, compared with 30 or 31 days for the original data.

We can prevent this distortion by randomly imputing the day. A simple procedure is to choose a day at random from the 28, 29, 30, or 31 days of the month with the missing day. But exactly what is the correct number of days? This tip was first drafted 10 October 2011. For most people, 31 is an easy answer for the correct number of days in October. It is easy for the authors because early in their educations, they memorized a little poem about the months that included the exceptions for February and leap years.

But how do we get Stata to do this calculation? What is crucial here is that Stata already knows the calendar, so it does not need to be instructed about month lengths or leap years. Instead, the trick is to recognize that the last day of the current month—in this example, 31—is just one day before the first day of the next month. Thus we can get this by calculating

```
. generate modays = day(dofm(monthlydate + 1) - 1)
```

`monthlydate + 1` is the next month; `dofm()` yields the first day of that month as a daily date and subtracting 1 gives us the last day of this month as a daily date; finally, `day()` yields the day of the month of that last day. This works regardless of leap years or whether the next month is also in the next year.

Then we can randomly impute a day in the month by calculating

```
. generate imputedday = ceil(modays * runiform())
```

However we impute the date, we can now fix the missing dates:

```
. replace dailydate = date(reported + string(imputedday), "YMD")
> if missing(dailydate)
```

We leave behind the variable `imputedday` in the dataset, because its nonmissing values indicate what was imputed and where each was imputed in the data.

The assumption entailed by random sampling from a discrete uniform distribution is clearly that any day of the month has the same probability. If that seems unlikely— for example, because of seasonality—some other procedure may be advisable. In the same vein, it will be recognized that the imputation does not use any other information that might be available. In the original medical example, that could include other observations on the same patient. Although these and other complications may be problematic in practice, we leave them for another day.

# References

Little, R. J. A., and D. B. Rubin. 2002. *Statistical Analysis with Missing Data*. 2nd ed. Hoboken, NJ: Wiley.

Seastrom, M. M., S. Kaufman, and R. Lee. 2002. Appendix B: Evaluating the impact of imputations for item nonresponse. In *NCES Statistical Standards (NCES 2003601)*, ed. M. M. Seastrom. National Center for Education Statistics, Institute of Education Sciences. http://nces.ed.gov/statprog/2002/appendixb.asp.

The Stata Journal (2012)
**12**, Number 1, pp. 162–164

# Stata tip 106: With or without reference

Maarten L. Buis
Department of Sociology
Tübingen University
Tübingen, Germany
maarten.buis@uni-tuebingen.de

A convenient way to define a set of indicator variables (often called dummy variables) is to use Stata's factor-variable notation (see [U] **11.4.3 Factor variables**). In that case, the default is to leave one category out, the so-called reference category. However, the factor-variable notation also allows you to include an indicator variable for the reference category. This can provide a useful alternative representation of the same model. The estimation and interpretation of these models are best explained using examples, like the ones below.

```
. sysuse auto
(1978 Automobile Data)

. summarize weight if foreign == 0, meanonly

. generate c_weight = (weight - r(min))/2000

. label var c_weight "weight centered at lightest domestic car (short tons)"

. quietly regress price i.foreign c_weight

. estimates store a1

. quietly regress price ibn.foreign c_weight, noconstant

. estimates store b1

. estimates table a1 b1, b(%9.3g)
```

| Variable | a1 | b1 |
|---|---|---|
| foreign | | |
| 0 | (base) | 1034 |
| 1 | 3637 | 4671 |
| c_weight | 6641 | 6641 |
| _cons | 1034 | |

In this example, the average price of "domestic" (U.S.) cars is compared with the average price of "foreign" cars while controlling for the weight of the car. Model a1 uses the default method of using indicator variables. The results are interpreted as follows: the lightest domestic car costs on average \$1,034, and an equally light foreign car costs on average \$3,637 more. Model b1 includes both an indicator variable for foreign cars and an indicator variable for domestic cars. These results are interpreted as follows: the lightest domestic car costs on average \$1,034, and an equally light foreign car costs on average \$4,671.

It is useful to note three things about these results. First, these models are completely equivalent; they are just different ways of saying the same thing. Model a1 emphasizes the comparison of the categories, while model b1 emphasizes the levels in each category. Second, the two indicator variables in model b1 contain information that was present in the indicator variable and the constant in model a1. Thus in model b1, there is no information left to put in the constant. As a consequence, you must leave the constant out of model b1, which was done by adding the `noconstant` option. Third, it helps to center all variables in the model on some meaningful value. In this example, I centered the weight on the lightest domestic car. If I had not done that, then the prices in models a1 and b1 would refer to cars weighing 0 tons.

This trick can also be useful when you have interactions, as is shown in the example below. Model a2 uses the default parameterization, which leaves out the reference category for both `foreign` and `good`. Model b2 includes an indicator variable for the reference category of `foreign` but leaves the reference category out for `good`. Model c2 contains indicator variables for all reference categories.

```
. generate byte good = rep78 > 3 if rep78 < .
(5 missing values generated)
. quietly regress price i.foreign##i.good c_weight
. estimates store a2
. quietly regress price i.foreign ibn.foreign#i.good c_weight, noconstant
. estimates store b2
. quietly regress price ibn.foreign#ibn.good c_weight, noconstant
. estimates store c2
. estimates table a2 b2 c2, b(%9.3g)
```

| Variable | a2 | b2 | c2 |
|---|---|---|---|
| foreign | | | |
| 0 | (base) | 974 | |
| 1 | 3150 | 4124 | |
| | | | |
| good | | | |
| 1 | -251 | | |
| | | | |
| foreign#good | | | |
| 0 0 | (base) | (base) | 974 |
| 0 1 | (base) | -251 | 723 |
| 1 0 | (base) | (base) | 4124 |
| 1 1 | 708 | 457 | 4581 |
| | | | |
| c_weight | 6711 | 6711 | 6711 |
| _cons | 974 | | |

Consider models a2 and b2. Model a2 says that a bad, light domestic car will cost \$974, while a similar foreign car will cost \$3,150 more. Model b2 says that the bad, light domestic car costs \$974, while a similar foreign car will cost \$4,124. Model a2 says that good cars are \$251 cheaper if they are domestic cars, while the effect of being a good car increases by \$708 if the car is foreign. Model b2 says that the effect of being a good car is −\$251 for domestic cars and \$457 for foreign cars.

Consider models b2 and c2. Model c2 says that bad, light domestic cars cost \$974, while good, light domestic cars cost \$723. Model b2 says that bad, light domestic cars cost \$974, while good domestic cars cost \$251 less. Model c2 says that bad, light foreign cars cost \$4,124, while good, light foreign cars cost \$4,581. Model b2 says that bad, light foreign cars cost \$4,124, while good foreign cars cost \$457 more.

This trick is not limited to linear regression but can be applied to any model. For example, assume we are worried about the right-skewed nature of price and think that a log transformation would be better, but we want to continue making statements in terms of the average price and not in terms of the average log price. In that case, we can use `glm` with the `link(log)` option (see [R] **glm** and Cox et al. [2008]) or `poisson` (see [R] **poisson** and Wooldridge [2010]). An important difference with linear regression is that one interprets the exponentiated parameters, and these are interpreted in multiplicative terms rather than additive terms. Consider the example below. Model a3 says that a light domestic car will cost on average \$2,102, while a similar foreign car will cost 2.145 times as much. Model b3 says that a light domestic car will cost on average \$2,102, while a similar foreign car will cost on average \$4,509.

```
. quietly glm price i.foreign c_weight, link(log) eform
. estimates store a3
. quietly glm price ibn.foreign c_weight, noconstant link(log) eform
. estimates store b3
. estimates table a3 b3, b(%9.4g) eform
```

| Variable | a3 | b3 |
|---|---|---|
| foreign | | |
| 0 | (base) | 2102 |
| 1 | 2.145 | 4509 |
| c_weight | 3.516 | 3.516 |
| _cons | 2102 | |

# References

Cox, N. J., J. Warburton, A. Armstrong, and V. J. Holliday. 2008. Fitting concentration and load rating curves with generalized linear models. *Earth Surface Processes and Landforms* 33: 25–39.

Wooldridge, J. M. 2010. *Econometric Analysis of Cross Section and Panel Data*. 2nd ed. Cambridge, MA: MIT Press.

The Stata Journal (2012)
**12**, Number 1, pp. 165–166

# Stata tip 107: The baseline is now reported

Maarten L. Buis
Department of Sociology
Tübingen University
Tübingen, Germany
maarten.buis@uni-tuebingen.de

For a long time, Stata has had the capability to report exponentiated coefficients. Examples are the **or** option of **logit** and **ologit** (see [R] **logit** and [R] **ologit**); the **irr** option of **poisson**, **zip**, and **nbreg** (see [R] **poisson**, [R] **zip**, and [R] **nbreg**); and the **hr** and **tr** options of **streg** (see [ST] **streg**). Also see Newson (2003) and Buis (2010) for details. These exponentiated coefficients can be interpreted as odds ratios, incidence-rate ratios, hazard ratios, or time ratios. However, until Stata 12 the baseline odds, incidence rate, hazard, or time—that is, the exponentiated constant—was not reported. That was unfortunate because this baseline can be helpful for evaluating the size of the effects, and it provides a convenient way of discussing the exact interpretation of the coefficients.

As of Stata 12, this omission has been redressed. The usefulness of the baseline value and a couple of caveats are illustrated using the example below.

```
. sysuse nlsw88
(NLSW, 1988 extract)

. generate c_grade = grade - 12
(2 missing values generated)

. generate high_occ = occupation < 3 if occupation < .
(9 missing values generated)

. logit union c_grade i.high_occ, or nolog

Logistic regression                             Number of obs   =       1867
                                                LR chi2(2)      =      49.44
                                                Prob > chi2     =     0.0000
Log likelihood = -1016.5579                     Pseudo R2       =     0.0237

------------------------------------------------------------------------------
       union |  Odds Ratio   Std. Err.      z    P>|z|     [95% Conf. Interval]
-------------+----------------------------------------------------------------
     c_grade |   1.123325   .0248694     5.25   0.000     1.075624    1.173141
 1.high_occ |   .4651723   .0644307    -5.53   0.000     .3545803    .6102575
       _cons |   .3358115     .02213   -16.56   0.000     .2951218    .3821112
------------------------------------------------------------------------------
```

Odds ratios have a bad reputation for being hard to interpret. Part of the problem is that many people are not used to working with odds. Researchers rarely frequent race tracks or betting shops. Starting the results section of an article with interpreting the baseline odds is a nice way to remind the readers of the correct interpretation. This trick works well because it fits naturally within the normal format of an academic article. In this case, we expect to find 0.34 union members for every nonmember within the group of respondents that has 12 years of education (c_grade = 0) and a lower occupation (high_occ = 0). The odds ratios tell us that the odds increases by a factor of 1.12 or

12% $[(1.12 - 1) \times 100\% = 12\%]$ for every additional year of education, while the odds decreases 53% $[(0.47 - 1) \times 100\% = -53\%]$ when the respondent has a high occupation. Reporting the baseline odds in the results section of a paper allows you to translate the abstract concept of odds to the concrete situation that is being studied; in this case, it allows you to translate "the number of successes per failure" to "the number of union members per nonmember".[1]

A 53% decrease in the odds of being a union member sounds like a large effect. However, we can get a better understanding of the size of this effect by comparing it with the baseline odds. In this case, the odds changes from 0.34 union members per nonunion member for respondents with lower occupations to 0.16 $(0.47 \times 0.34 = 0.16)$ union members per nonmember, which is a substantively meaningful change. But what if being a union member was very rare? For example, assume that the baseline odds was 0.001 union member for every nonunion member. In that case, the odds would change from 0.001 to 0.00047 union members per nonmember when a respondent obtained a high occupation, which does not sound nearly as impressive as a change of $-53\%$. So the baseline value can play an important role in evaluating how large an effect is.

There are, however, a couple of things you need to consider when interpreting these baseline values. First, the baseline value is the value when all explanatory variables are 0. So to get a meaningful baseline value, you need to make sure the value 0 is meaningful for all explanatory variables. In the example above, I did so by centering the variable `grade` at 12 years of education (obtaining high school). Second, the practice of reporting $p$-values or assigning stars to significant parameters needs a bit of thought in the case of baseline values. Stata automatically reports the results of the test of the null hypothesis that the coefficient is 0, and thus the exponentiated coefficient is 1. In the example, that would mean that the null hypothesis for the baseline odds is that there is 1 union member for every nonmember; that is, the probability of being a union member is 50%.

# References

Agresti, A. 2007. *An Introduction to Categorical Data Analysis*. 2nd ed. Hoboken, NJ: Wiley.

Buis, M. 2010. Stata tip 87: Interpretation of interactions in nonlinear models. *Stata Journal* 10: 305–308.

Fienberg, S. E. 2007. *The Analysis of Cross-Classified Categorical Data*. 2nd ed. New York: Springer.

Newson, R. 2003. Stata tip 1: The eform() option of regress. *Stata Journal* 3: 445.

---

1. More complete discussions of odds and odds ratios can be found in the textbooks by Fienberg (2007) and Agresti (2007).

The Stata Journal (2012)
**12**, Number 2, pp. 342–344

# Stata tip 108: On adding and constraining

Maarten L. Buis
Department of Sociology
Tübingen University
Tübingen, Germany
maarten.buis@uni-tuebingen.de

In many estimation commands, the `constraint` command (see [R] **constraint**) can impose linear constraints. The most common of these is the constraint that two or more regression coefficients are equal. A sometimes useful characteristic of models with that constraint is that they are equivalent to a model that includes the sum of the variables that are constrained. Consider the relevant part of a regression equation:

$$\beta_1 x_1 + \beta_2 x_2$$

If we constrain the effects of $x_1$ and $x_2$ to be equal, then we can replace $\beta_1$ and $\beta_2$ with $\beta$:

$$\beta x_1 + \beta x_2 = \beta(x_1 + x_2)$$

One situation where this characteristic can be useful occurs when you have created a variable by adding several variables and you wonder whether that was a good idea. In the example below, there are three variables on the degree of trust a respondent has in the executive, legislative, and judicial branches of the U.S. federal government: `confed`, `conlegis`, and `conjudge`, respectively. These can take the values 0 (hardly any trust), 1 (only some trust), or 2 (a great deal of trust). I think that these three variables say something about the trust in the federal government, and I created a single variable that captures that, `congov`, which I use to predict whether a respondent voted for Barack Obama in the 2008 U.S. presidential election. This results in model `sum1`.

If I want to check whether adding these three confidence measures was a good idea, I can use the fact that adding variables is equivalent to constraining their effects to be equal. So you can operationalize the rather vague idea "adding these variables is a good idea" to the testable statement "the effects of these three variables are the same". As a check, I first fit a model that constrains the effects to be equal. This is model `constr1`, and as expected, the resulting coefficients, standard errors, and log likelihood are exactly the same. I then fit a model with the three confidence variables without constraint, `unconstr1`. The resulting coefficients are very different from one another: the effects do not even have the same sign.[1] A likelihood-ratio test also rejects the hypothesis that these variables have the same effect on voting for Obama. So adding the sum of the three confidence measures was not a good idea in this case.

---

1. These are odds ratios, so the sign is determined by whether the ratio is larger or smaller than 1.

```
. use gss10
(extract from the 2010 General Social Survey)
. generate byte congov = confed + conlegis + conjudge
(463 missing values generated)
. quietly logit obama congov, or nolog
. estimates store sum1
. constraint 1 confed = conlegis
. constraint 2 confed = conjudge
. quietly logit obama confed conlegis conjudge, or constraint(1 2) nolog
. estimates store constr1
. quietly logit obama confed conlegis conjudge, or nolog
. estimates store unconstr1
. estimates table sum1 constr1 unconstr1, stats(ll N) eform b(%9.3g) se(%9.3g)
> stfmt(%9.4g)
```

| Variable | sum1 | constr1 | unconstr1 |
|---|---|---|---|
| congov | 1.62 | | |
| | .11 | | |
| confed | | 1.62 | 3.47 |
| | | .11 | .576 |
| conlegis | | 1.62 | 1.69 |
| | | .11 | .305 |
| conjudge | | 1.62 | .674 |
| | | .11 | .107 |
| _cons | .461 | .461 | .689 |
| | .0833 | .0833 | .134 |
| ll | -347.8 | -347.8 | -324.9 |
| N | 557 | 557 | 557 |

legend: b/se

```
. lrtest constr1 unconstr1
Likelihood-ratio test                                    LR chi2(2)  =      45.77
(Assumption: constr1 nested in unconstr1)                Prob > chi2 =     0.0000
```

Another situation where this characteristic can be useful occurs when you have two or more ordinal or categorical variables that you want to combine. Consider the example below. In that example, I want to treat education as an ordinal variable, and I want to see the effect of "family educational background" on the educational attainment of the children.

I think of family educational background as some sort of sum of the father's and mother's educations, but how do I create a sum of two ordinal variables? That is hard, but it is easy to consider the equivalent model that constrains the effects of father's education to be equal to the effects of mother's education. In this example, the effects of mother's and father's educations are fairly similar, and the test of the hypothesis that they are equal cannot be rejected (compare unconstr2 with constr2). It also shows that constraining effects to be the same is equivalent to adding the sums of the indicator variables (compare constr2 with sum2).

```
. quietly ologit degree i.madeg i.padeg, or nolog

. estimates store unconstr2

. constraint 1 1.madeg = 1.padeg

. constraint 2 2.madeg = 2.padeg

. quietly ologit degree i.madeg i.padeg, or constraint(1 2) nolog

. estimates store constr2

. generate byte p_hs = 1.madeg + 1.padeg
(329 missing values generated)

. generate byte p_mths = 2.madeg + 2.padeg
(329 missing values generated)

. quietly ologit degree p_hs p_mths, or nolog

. estimates store sum2

. estimates table unconstr2 constr2 sum2, stats(ll N) eform b(%9.3g) se(%9.3g)
> stfmt(%9.4g) keep(degree:)
```

| Variable | unconstr2 | constr2 | sum2 |
|---|---|---|---|
| madeg |  |  |  |
| 1 | 2.5 | 2.17 |  |
|  | .449 | .199 |  |
| 2 | 4.89 | 5.51 |  |
|  | 1.26 | .691 |  |
|  |  |  |  |
| padeg |  |  |  |
| 1 | 1.88 | 2.17 |  |
|  | .322 | .199 |  |
| 2 | 6.08 | 5.51 |  |
|  | 1.46 | .691 |  |
|  |  |  |  |
| p_hs |  |  | 2.17 |
|  |  |  | .199 |
| p_mths |  |  | 5.51 |
|  |  |  | .691 |
|  |  |  |  |
| ll | -799.1 | -800.5 | -800.5 |
| N | 972 | 972 | 972 |

legend: b/se

```
. lrtest unconstr2 constr2
```

| Likelihood-ratio test | LR chi2(2) = | 2.79 |
|---|---|---|
| (Assumption: constr2 nested in unconstr2) | Prob > chi2 = | 0.2484 |

The Stata Journal (2012)
**12**, Number 2, pp. 345–346

# Stata tip 109: How to combine variables with missing values

Peter A. Lachenbruch
Oregon State University (retired)
Corvallis, OR
peter.lachenbruch@oregonstate.edu

A common problem in data management is combining two or more variables with missing values to get a single variable with as many nonmissing values as possible. Typically, this problem arises when what should be the same variable has been named differently in different datasets. Stata's treatment of missing values means that the combination needs a little care, although there are several quite easy solutions.

To make this situation concrete, consider a recent experience of mine. Two datasets were created. A variable `resp_ill` was created in one dataset, and the corresponding variable `moxResp` was created in the other. After appending the data with the `append` command, there were missing values on `resp_ill` if the data came from the second dataset, and there were missing values on `moxResp` if the data came from the first dataset. The total number of observations is 782. Here is a short listing of some of the observations:

```
. list if inrange(_n, 1, 5) | inrange(_n, 340, 344), separator(0)
```

|      | resp_ill | moxResp |
|------|----------|---------|
| 1.   | no       | .       |
| 2.   | no       | .       |
| 3.   | no       | .       |
| 4.   | no       | .       |
| 5.   | no       | .       |
| 340. | .        | no      |
| 341. | .        | no      |
| 342. | .        | no      |
| 343. | .        | no      |
| 344. | .        | yes     |

The variables are indicators of whether the patient had a respiratory illness. They have been labeled as "no" for 0 and "yes" for 1.

It is easy to see or to say that the variables should be renamed (with the `rename` command) for consistency. I could have done that in one of the original datasets and then repeated the `append` command. In practice, it is often quicker to fix the problem in the current dataset. Sometimes, maintaining dataset integrity is sufficiently important to rule out the `rename` solution altogether.

Experimentation will confirm that adding the two variables is not the answer, because the missing values are propagated: for Stata, nonmissing plus missing is still missing. Doing that with a `replace` of one variable rather than a `generate` would be an even worse idea because any original nonmissing values would be lost in the variable that was replaced.

What we want is to ignore the missings, and there are several ways to do that. One explicit solution is

```
. generate newind = min(resp_ill, moxResp)
```

The pairwise minimum will be whichever nonmissing value is present. What is less intuitive is that the function `max()` will work in the same way because it follows the same principle of ignoring missings to the extent possible, just as, say, `summarize` returns the maximum nonmissing value and not missings. So even though Stata has a general rule that missings count higher than nonmissings, in the case of the `max()` function, that rule is trumped by the principle that missings are ignored as far as possible.

Another solution is

```
. generate newind = cond(missing(resp_ill), moxResp, resp_ill)
```

In this particular case, it was known that the missings in one variable corresponded to the nonmissings in the other. In other circumstances, it would be prudent to check whether there were observations with nonmissing values on both variables, say, by typing

```
. list if var1 < . & var2 < .
```

Another solution that you may know is to use `egen`'s `rowtotal()` function, which by default ignores missings in forming a row total and so gives a single nonmissing result for one missing and one nonmissing argument. Using `egen` can be convenient, especially if you have more than two variables to combine, but it does impose an extra overhead because there is more code for Stata to interpret. For one-off tasks and small- or moderate-size datasets, you would have to strain to detect that overhead, but for repeated data management and large datasets, using `min()` or `max()` is preferable.

The Stata Journal (2012)
**12**, Number 2, pp. 347–351

# Stata tip 110: How to get the optimal k-means cluster solution

Anna Makles
Schumpeter School of Business and Economics
University of Wuppertal
Wuppertal, Germany
makles@statistik.uni-wuppertal.de

The $k$-means cluster algorithm is a well-known partitional clustering method but is also widely used as an iterative or exploratory clustering method within unsupervised learning procedures (Hastie, Tibshirani, and Friedman 2009, chap. 14). When the number of clusters is unknown, several $k$-means solutions with different numbers of groups $k$ ($k = 1, \ldots, K$) are computed and compared. To detect the clustering with the *optimal* number of groups $k^*$ from the set of $K$ solutions, we typically use a scree plot and search for a kink in the curve generated from the within sum of squares (WSS) or its logarithm [$\log(\text{WSS})$] for all cluster solutions. Another criterion for detecting the optimal number of clusters is the $\eta^2$ coefficient, which is quite similar to the $R^2$, or the proportional reduction of error (PRE) coefficient (Schwarz 2008, 72):

$$\eta_k^2 = 1 - \frac{\text{WSS}(k)}{\text{WSS}(1)} = 1 - \frac{\text{WSS}(k)}{\text{TSS}} \quad \forall\, k \in K$$

$$\text{PRE}_k = \frac{\text{WSS}(k-1) - \text{WSS}(k)}{\text{WSS}(k-1)} \quad \forall\, k \geq 2$$

Here $\text{WSS}(k)$ [$\text{WSS}(k-1)$] is the WSS for cluster solution $k$ ($k-1$), and $\text{WSS}(1)$ is the WSS for cluster solution $k = 1$, that is, for the nonclustered data. $\eta_k^2$ measures the proportional reduction of the WSS for each cluster solution $k$ compared with the total sum of squares (TSS). In contrast, $\text{PRE}_k$ illustrates the proportional reduction of the WSS for cluster solution $k$ compared with the previous solution with $k-1$ clusters.

Because the `cluster kmeans` command does not store any results in `e()`, we must use the same trick as in the `cluster stop` ado-file for hierarchical clustering to gather the information on the WSS for different cluster solutions. The following example uses 20 different cluster solutions, $k = 1, \ldots, 20$, and `physed.dta`, which measures different characteristics of 80 students and is discussed in [MV] **cluster kmeans and kmedians**. The dataset is available at

```
. use http://www.stata-press.com/data/r12/physed
```

After the variables `flexibility`, `speed`, and `strength` are standardized by typing

```
. local list1 "flex speed strength"
. foreach v of varlist `list1´ {
  2. egen z_`v´ = std(`v´)
  3. }
```

we calculate 20 cluster solutions with random starting points and store the results in
name(*clname*):

```
. local list2 "z_flex z_speed z_strength"
. forvalues k = 1(1)20 {
  2. cluster kmeans `list2´, k(`k´) start(random(123)) name(cs`k´)
  3. }
```

To gather the WSS of each cluster solution cs`k', we calculate an ANOVA using the
anova command, where cs`k' is the cluster variable. anova stores the residual sum of
squares for the chosen variable within the defined groups in cs`k' in e(rss), which is
exactly the same as the variable's sum of squares within the clusters. To collect the
information on all cluster solutions, we generate a 20 × 5 matrix to store the WSS, its
logarithm, and both coefficients for every cluster solution $k$.

```
. * WSS matrix
. matrix WSS = J(20,5,.)
. matrix colnames WSS = k WSS log(WSS) eta-squared PRE
. * WSS for each clustering
. forvalues k = 1(1)20 {
  2. scalar ws`k´ = 0
  3.   foreach v of varlist `list2´ {
  4.   quietly anova `v´ cs`k´
  5.   scalar ws`k´ = ws`k´ + e(rss)
  6.   }
  7. matrix WSS[`k´, 1] = `k´
  8. matrix WSS[`k´, 2] = ws`k´
  9. matrix WSS[`k´, 3] = log(ws`k´)
 10. matrix WSS[`k´, 4] = 1 - ws`k´/WSS[1,2]
 11. matrix WSS[`k´, 5] = (WSS[`k´-1,2] - ws`k´)/WSS[`k´-1,2]
 12. }
```

Finally, we use the columns of the output matrix WSS and the _matplot command
to produce plots of the calculated statistics.

```
. matrix list WSS

WSS[20,5]
              k         WSS    log(WSS)  eta-squared          PRE
    r1        1         237   5.4680601            0            .
    r2        2   89.351871   4.4925822    .62298789    .62298789
    r3        3   56.208349   4.0290653    .76283397     .3709326
    r4        4   16.471059   2.8016049    .93050186    .70696419
    r5        5   13.823239   2.6263512     .9416741    .16075591
    r6        6   12.737676   2.5445642    .94625453    .07853172
    (output omitted )
. local squared = char(178)
. _matplot WSS, columns(2 1) connect(l) xlabel(#10) name(plot1, replace) nodraw
> noname
. _matplot WSS, columns(3 1) connect(l) xlabel(#10) name(plot2, replace) nodraw
> noname
. _matplot WSS, columns(4 1) connect(l) xlabel(#10) name(plot3, replace) nodraw
> noname ytitle({&eta}`squared´)
```

```
  . _matplot WSS, columns(5 1) connect(l) xlabel(#10) name(plot4, replace) nodraw
  > noname
  (1 points have missing coordinates)
  . graph combine plot1 plot2 plot3 plot4, name(plot1to4, replace)
```

The results indicate clustering with $k = 4$ to be the optimal solution. At $k = 4$, there is a kink in the WSS and log(WSS), respectively. $\eta_4^2$ points to a reduction of the WSS by 93% and $PRE_4$ to a reduction of about 71% compared with the $k = 3$ solution. However, the reduction in WSS is negligible for $k > 4$.

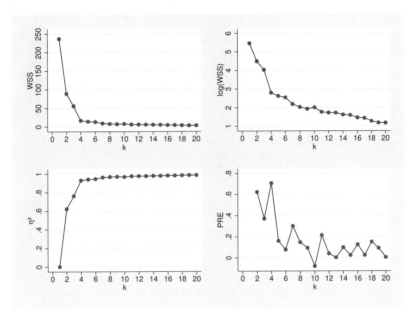

Figure 1. WSS, log(WSS), $\eta^2$, and PRE for all $K$ cluster solutions

In figure 2, we see a scatterplot matrix of the standardized variables for the four-cluster solution, which indicates the four distinct groups of students.

```
  . graph matrix z_flex z_speed z_strength, msym(i) mlab(cs4) mlabpos(0)
  > name(matrixplot, replace)
```

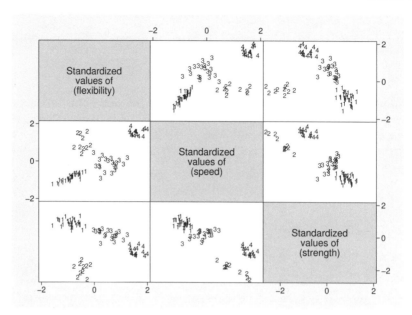

Figure 2. Scatterplot matrix of the standardized variables for the four-cluster solution

Although the results seem quite clear, this is not always the case. The results of a traditional $k$-means algorithm always depend on the chosen initialization (that is, the initial cluster centers) and, of course, the data.

Figure 3 again shows results for `physed.dta` but for 50 different starting points. Here our optimal solution with four clusters occurs 37 times (75%). Ten (20%) results point to the five-cluster solution to be the optimal number of groups. Hence, depending on the initialization, *natural* clusters may be divided into subgroups, or sometimes no kink is even visible. The best way to evaluate the chosen solution is therefore to repeat the clustering several times with different starting points and then compare the different solutions as done here.

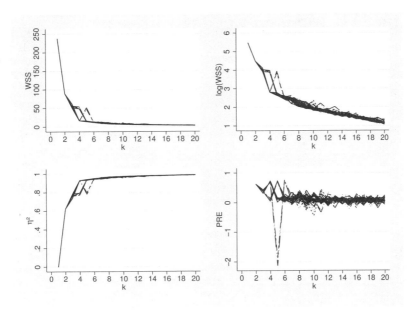

Figure 3. Fifty different WSS, log(WSS), $\eta^2$, and PRE curves for $K = 20$

# References

Hastie, T., R. J. Tibshirani, and J. Friedman. 2009. *The Elements of Statistical Learning: Data Mining, Inference, and Prediction*. 2nd ed. New York: Springer.

Schwarz, A. 2008. *Lokale Scoring-Modelle*. Lohmar, Germany: Eul Verlag.

The Stata Journal (2012)
**12**, Number 3, pp. 565–569

290

# Stata tip 111: More on working with weeks

Nicholas J. Cox
Department of Geography
Durham University
Durham, UK
n.j.cox@durham.ac.uk

## 1 Introduction

Researchers in several fields work with weekly data. The most prominent examples as far as Stata is concerned come from business, economics, and finance. In particular, the Federal Reserve Bank of St. Louis offers a repository of many economic time series, some of which are weekly. See http://research.stlouisfed.org/fred2/ and documentation on the Stata command `freduse` (Drukker 2006). The M2 money stock for weeks ending on Monday and the St. Louis Financial Stress Index for weeks ending on Friday are specific examples.

Like any kind of time-series data, weekly data vary in character. In some cases, the data are totals, such as sales totals. It seems typical in such cases that any within-week variations (including zeros for days when a business is closed and no transactions occur) are not considered interesting or important. In other cases, the data are weekly averages, but weekly cycles are not reported on similar grounds. In yet other cases, weekly values are point values: the underlying variable is continuously or discretely varying but just happens to be reported once a week. These differences do not affect how weekly data are held within Stata, but they do affect interpretations of analyses and how weekly data should be aggregated to longer intervals.

Some advice on dealing with such weekly data was given in a previous Stata tip (Cox 2010). The main purposes of this tip are to further discuss how to convert data presented in yearly and weekly form to daily dates and how to aggregate such data to months or longer intervals.

Kit Baum, Kevin Crow, and David Drukker made helpful comments on the issues discussed here.

## 2 Previous tip

The previous tip emphasized that

1. Stata's own definition of weekly dates is idiosyncratic, hinging on the idea that January 1 is always the start of the first week in any given year and that there are always precisely 52 weeks in any year. Thus week 52 is either 8 or 9 days long, depending on whether a year is a leap year. Defined this way, weeks always fit into years.

2. any week defined in the more common way as being always 7 days long and so starting, or equivalently finishing, on a particular day of the week can always be defined by that particular day of the week.

3. Stata's `dow()` function is useful for manipulations. It returns 0 for Sundays, 1 for Mondays, and so forth until 6 for Saturdays.

# 3   Converting year–week data to Stata daily dates

Data for the St. Louis Financial Stress Index can be read into Stata by typing

```
. freduse STLFSI
```

This is an example of easy-to-use weekly data referred to daily dates that are 7 days apart. Such data may be declared time-series data by specifying `delta(7)` with `tsset` or `xtset`. It would not be a good idea to try to declare the dates as Stata weekly dates or even to try to convert them to Stata weekly dates. Weeks defined by beginning or ending days may span two different years, and it is entirely possible for 53 weeks to occur in some years. However, Stata's weekly dates are defined too rigidly to allow either variation.

Yet users often report that their weekly data arrive in years and weeks, such as year 2012, week 31. We need a way of converting that kind of data to equivalent daily dates. We need only the day of the week considered to characterize each week. Whether that day is at the beginning or end of the week (or even in the middle) is for the moment immaterial.

Here is an example. We have a composite variable `yearweek` specifying year and week. We will use Friday as the key day as a first example.

```
. list yearweek
```

|       | yearweek |
|-------|----------|
| 1.    | 201148   |
| 2.    | 201149   |
| 3.    | 201150   |
| 4.    | 201151   |
| 5.    | 201152   |
| 6.    | 201201   |
| 7.    | 201202   |
| 8.    | 201203   |
| 9.    | 201204   |
| 10.   | 201205   |

If such a variable were a string variable, we could split the components by typing

```
. generate year = real(substr(yearweek, 1, 4))
. generate week = real(substr(yearweek, -2, 2))
```

If such a variable were a numeric variable, we could do it this way:

```
. generate year = floor(yearweek/100)
. generate week = mod(yearweek, 100)
```

Naturally, if we started with separate variables for year and week, neither conversion would be necessary.

The calculation pivots on finding the first Friday in each year. Once we have done that, every other Friday in that year is just a multiple of seven days later. It is usually convenient, if a little wasteful in storage, to start by defining a variable that is the day of the week for each January 1. (If, as an exception, data all lie within a single year, then there is a single value to deal with, and we need not use a variable.)

```
. generate jan1 = dow(mdy(1,1,year))
```

Let's use a local macro to hold the day of the week we want, if only to expose the generality of what we are doing. `dow()` returns 5 for Fridays.

```
. local wanted = 5
```

We can be sure that the first Friday of the year is within the first 7 days of the year. If the first day of the year is in fact Sunday to Friday, then the first Friday is respectively January 6, 5, 4, 3, 2, or 1. If the first day of the year is in fact Saturday, we have to wait until January 7. If we were identifying Mondays, the sequence would be another rotation of the integers 7, ..., 1 (namely, January 2, 1, 7, 6, 5, 4, 3). Experiment or deduction yields the general rule for the first day of interest within the year as

```
. generate weekdates =
> cond(jan1 <= `wanted', 1 - jan1 + `wanted', 8 - jan1 + `wanted')
```

That is, for another day of the week other than Friday, the definition of **wanted** would change, but the command above would still be valid. See Kantor and Cox (2005) for a tutorial on the `cond()` function.

For the weeks actually present in our data, we need to do the following:

```
. replace weekdates = weekdates +  7 * (week - 1)
(9 real changes made)
```

These are, or rather should be, days of the year somewhere between 1 and 365 or 366, depending on whether each year in question is a leap year. We do not have to agonize about identifying leap years because there is an easy way to identify whether December 31 is day 365 or 366. If the day of the year calculated is more than that, there is an error in the dataset, which we should want to know about.

```
. replace weekdates = . if weekdates > doy(mdy(12, 31, year))
(0 real changes made)
```

Then the conversion to daily dates is straightforward:

```
. generate date = mdy(1,1,year) + weekdates - 1
. format date %td
```

These are daily dates, but as I explained earlier, we can get the best of both worlds by specifying `delta(7)` with `tsset` or `xtset`.

We could now type

```
. drop jan1
```

Many problems are now easier. For example, suppose we wanted to change the characterization of weeks by tagging in terms of first days rather than last days or even by tagging by a middle day (which might seem appropriate for graphics). All such changes are just adding or subtracting a fixed number of days. In what follows, we will take the dates in our example as defining the end of each week.

## 4    Aggregating weeks to longer intervals

Suppose we want to aggregate weekly values to longer intervals such as months. We will focus on months, because further aggregation of months to quarters, half-years, or years is easy. In practice, aggregation usually entails averaging or summing values, according to the nature of the variable. Aggregation may follow creation of a monthly date by using the standard function `mofd()`, but how to treat weeks that span months remains an issue.

A partial week ends on or before the 6th of any month or begins 6 or fewer days before the end of any month, which means 7 or fewer days before the beginning of any month. If we work with weeks defined by their beginning days, then the end of a month could be day 28, 29, 30, or 31, depending on the month and whether a year is a leap year. In practice, it can be easier to consider the end of the month as being also the day before the first day of the next month, as explained in Samuels and Cox (2012). However, in practice, working with weeks defined by their ending days is even easier, and we just explained how to calculate such a variable if one does not already exist.

Let's fake some data for our example:

```
. set seed 2803
. generate y = ceil(42 * sqrt(runiform()))
```

Then `mofd()` gives us the monthly date corresponding to each daily date:

```
. generate month = mofd(date)
. format month %tm
```

If the data were spot or point data for the dates specified, it is simplest, and may even be best, just to average all data for each month. How might we deal otherwise with the problem of spanning weeks?

First, we will generate the lesser of the day of the month and 7. This will be between 1 and 6 for weeks that end on those days of the month and 7 otherwise. Thus we have a variable that tags those weeks spanning two months and indicates how many days fall in each of the two months (because if $x$ days fall in the second month, $7 - x$ days fall in the first).

```
. generate length = min(day(date), 7)
```

Keeping track of the number of observations in the original dataset, we now need to split each spanning week between two months. The trick is to expand the data first with the `expand` command and then collapse the data with the `collapse` command.

```
. local N = _N
. expand 2 if length < 7
(3 observations created)
. replace month = month - 1 if _n > `N'
(3 real changes made)
. replace length = 7 - length if _n > `N'
(3 real changes made)
. collapse (mean) y (count) days=length [fw=length], by(month)
. list
```

|     | month   | y        | days |
|-----|---------|----------|------|
| 1.  | 2011m11 | 32       | 5    |
| 2.  | 2011m12 | 23.09678 | 31   |
| 3.  | 2012m1  | 31.12903 | 31   |
| 4.  | 2012m2  | 42       | 3    |

In this example, the weighted mean of the outcome and the number of the days with observations are produced. `collapse` can produce many other reductions of raw data. `contract` is an alternative for some problems.

# References

Cox, N. J. 2010. Stata tip 68: Week assumptions. *Stata Journal* 10: 682–685.

Drukker, D. M. 2006. Importing Federal Reserve economic data. *Stata Journal* 6: 384–386.

Kantor, D., and N. J. Cox. 2005. Depending on conditions: A tutorial on the cond() function. *Stata Journal* 5: 413–420.

Samuels, S. J., and N. J. Cox. 2012. Stata tip 105: Daily dates with missing days. *Stata Journal* 12: 159–161.

The Stata Journal (2012)
**12**, Number 4, pp. 759–760

# Stata tip 112: Where did my p-values go? (Part 2)

Maarten L. Buis
Wissenschaftszentrum Berlin für Sozialforschung (WZB)
Berlin, Germany
maarten.buis@wzb.eu

In a previous Stata tip (Buis 2007), I discussed how to recover $t$ statistics, $p$-values, and confidence intervals for regression parameters by using the results that are returned by an estimation command. In this tip, I continue that discussion by showing how $p$-values can be recovered for other tests that are sometimes displayed by estimation commands. For example, consider a linear regression as estimated by `regress` (see [R] **regress**). It displays the results of an $F$ test of the hypothesis that all coefficients except the constant are equal to 0. However, `regress` only returns the $F$ statistic (`e(F)`), the number of model degrees of freedom (`e(df_m)`), and the number of residual degrees of freedom (`e(df_r)`); it does not return the $p$-value. If you need the $p$-value, you can use the function `Ftail()` to look up the appropriate $p$-value, as illustrated below.

```
. sysuse auto
(1978 Automobile Data)

. regress price mpg i.rep78
```

| Source | SS | df | MS |  |  |
|---|---|---|---|---|---|
| Model | 149020603 | 5 | 29804120.7 |  |  |
| Residual | 427776355 | 63 | 6790100.88 |  |  |
| Total | 576796959 | 68 | 8482308.22 |  |  |

|  | Number of obs = | 69 |
|---|---|---|
|  | F( 5, 63) = | 4.39 |
|  | Prob > F = | 0.0017 |
|  | R-squared = | 0.2584 |
|  | Adj R-squared = | 0.1995 |
|  | Root MSE = | 2605.8 |

| price | Coef. | Std. Err. | t | P>\|t\| | [95% Conf. Interval] | |
|---|---|---|---|---|---|---|
| mpg | -280.2615 | 61.57666 | -4.55 | 0.000 | -403.3126 | -157.2103 |
| rep78 |  |  |  |  |  |  |
| 2 | 877.6347 | 2063.285 | 0.43 | 0.672 | -3245.51 | 5000.78 |
| 3 | 1425.657 | 1905.438 | 0.75 | 0.457 | -2382.057 | 5233.371 |
| 4 | 1693.841 | 1942.669 | 0.87 | 0.387 | -2188.274 | 5575.956 |
| 5 | 3131.982 | 2041.049 | 1.53 | 0.130 | -946.7282 | 7210.693 |
| _cons | 10449.99 | 2251.041 | 4.64 | 0.000 | 5951.646 | 14948.34 |

```
. display Ftail(e(df_m), e(df_r), e(F))
.00171678
```

Often such additional tests are based on the chi-squared distribution. In that case, we can use the `chi2tail()` function to recover the $p$-value. An example is given below. In this example, the test statistic is returned in `e(chi2_c)`. The number of degrees of freedom for this test is not returned by `biprobit` (see [R] **biprobit**), but we know that in this case the number of degrees of freedom has to be 1.

```
. webuse school
. biprobit private vote logptax loginc years, nolog
Bivariate probit regression                        Number of obs    =        95
                                                   Wald chi2(6)     =      9.59
Log likelihood = -89.254028                        Prob > chi2      =    0.1431
```

|              | Coef.      | Std. Err. | z     | P>\|z\| | [95% Conf. | Interval]  |
|--------------|------------|-----------|-------|---------|------------|------------|
| **private**  |            |           |       |         |            |            |
| logptax      | -.1066962  | .6669782  | -0.16 | 0.873   | -1.413949  | 1.200557   |
| loginc       | .3762037   | .5306484  | 0.71  | 0.478   | -.663848   | 1.416255   |
| years        | -.0118884  | .0256778  | -0.46 | 0.643   | -.0622159  | .0384391   |
| _cons        | -4.184694  | 4.837817  | -0.86 | 0.387   | -13.66664  | 5.297253   |
| **vote**     |            |           |       |         |            |            |
| logptax      | -1.288707  | .5752266  | -2.24 | 0.025   | -2.416131  | -.1612839  |
| loginc       | .998286    | .4403565  | 2.27  | 0.023   | .1352031   | 1.861369   |
| years        | -.0168561  | .0147834  | -1.14 | 0.254   | -.0458309  | .0121188   |
| _cons        | -.5360573  | 4.068509  | -0.13 | 0.895   | -8.510188  | 7.438073   |
| /athrho      | -.2764525  | .2412099  | -1.15 | 0.252   | -.7492153  | .1963102   |
| rho          | -.2696186  | .2236753  |       |         | -.6346806  | .1938267   |

```
Likelihood-ratio test of rho=0:      chi2(1) =   1.38444     Prob > chi2 = 0.2393
. display chi2tail(1,e(chi2_c))
.23934684
```

# Reference

Buis, M. L. 2007. Stata tip 53: Where did my p-values go? *Stata Journal* 7: 584–586.

The Stata Journal (2012)
**12**, Number 4, pp. 761–764

# Stata tip 113: Changing a variable's format: What it does and does not mean

Nicholas J. Cox
Department of Geography
Durham University
Durham, UK
n.j.cox@durham.ac.uk

## 1    Introduction

Stata variables are all associated with a display format. Users can assign such a display format or work with a display format assigned by default. Thus suppose that you create new variables, for example,

```
. set obs 1
obs was 0, now 1
. generate mynum = 42
. generate mystr = "42"
```

Now type

```
. describe
Contains data
  obs:            1
  vars:           2
  size:           6

                  storage   display    value
variable name     type      format     label      variable label

mynum             float     %9.0g
mystr             str2      %9s

Sorted by:
     Note:  dataset has changed since last saved
```

You can see that `mynum` has been created as a `float` variable with display format `%9.0g` and that `mystr` has been created as a `str2` variable with display format `%9s`. If you do not like either of those formats, you can change them using the `format` command (see [D] **format**).

The problem addressed in this tip is that users are often puzzled over exactly what is meant by changing formats. Part of the problem behind such puzzlement may be linguistic. Sometimes, the term "format" is used vaguely or loosely, such as when "formatting" implies something like initial preparation or transformation of the data. At other times, the term may be used precisely but not in the sense of Stata's `format` command. Thus you may read of long or wide format in the sense of dataset shape or structure. Part of the solution to such puzzlement is thus also linguistic, to remember that "format" here means "display format".

# 2  Applying format

Consider `auto.dta`, in particular the `gear_ratio` variable.

```
. sysuse auto, clear
(1978 Automobile Data)
. codebook gear_ratio
```

| gear_ratio | | | | | Gear Ratio |
|---|---|---|---|---|---|

```
                 type:  numeric (float)
                range:  [2.19,3.89]                  units:  .01
        unique values:  36                      missing .:  0/74
                 mean:  3.01486
            std. dev:  .456287
          percentiles:        10%       25%       50%       75%       90%
                             2.43      2.73     2.955      3.37      3.72
```

`codebook` usefully reports that this variable has units (some say "resolution": for example, Murphy [1997]) of 0.01, meaning that values for this variable are given to 2 decimal places. Regardless of that, from the results of

```
. summarize gear_ratio
```

| Variable | Obs | Mean | Std. Dev. | Min | Max |
|---|---|---|---|---|---|
| gear_ratio | 74 | 3.014865 | .4562871 | 2.19 | 3.89 |

we can see that the default display of `summarize` shows many decimal places. The mean is reported to 6 decimal places and the standard deviation to 7, which is many more than are present in the original data. It can be asserted confidently that `summarize` shows far more minute detail than anyone can use or interpret. However, typing

```
. describe gear_ratio
```

| variable name | storage type | display format | value label | variable label |
|---|---|---|---|---|
| gear_ratio | float | %6.2f | | Gear Ratio |

shows that `gear_ratio` has a display format of `%6.2f`. If we wish `summarize` to honor that display format, we must specify the `format` option:

```
. summarize gear_ratio, format
```

| Variable | Obs | Mean | Std. Dev. | Min | Max |
|---|---|---|---|---|---|
| gear_ratio | 74 | 3.01 | 0.46 | 2.19 | 3.89 |

Now the results may be a little too Spartan for some tastes. One common suggestion is that a standard deviation can be reported a little more precisely than the original data. To get a more precise display, we can change the format of gear_ratio:

```
. format gear_ratio %6.3f
```

We see the result when we reissue the summarize command:

```
. summarize gear_ratio, format
    Variable |       Obs        Mean    Std. Dev.        Min         Max
-------------+--------------------------------------------------------
  gear_ratio |        74       3.015        0.456       2.190       3.890
```

This example is telling. The first thing to do is look for an option that changes the format of results. If there is not one, or it does not do what you want, then use format directly and reissue the command.

Moreover, it should be clear that format cannot introduce a third decimal place to the gear_ratio data that was never typed in originally, that is, at the time auto.dta was compiled. All format does is change how data and results are displayed. Even if the format is changed to one coarser than was entered, a detailed check will confirm that the data themselves are unchanged. One way to see this, left as an exercise for you, is to vary the format of some key variable and then check that the results from some interesting command remain completely identical. Use return list or ereturn list to get a high-resolution display.

# 3   Date formats

Dates are common in many problems but often do not arrive in exactly the right form for analysis. Hence users often want to convert dates as received to some other kind of date.

A common misconception is that changing the date format is the way to change one kind of date to another kind. That is wrong. Suppose we wish to change daily dates to monthly dates. As an experiment, set up a daily date variable:

```
. clear
. set obs 1
obs was 0, now 1
. generate mydate = d(8oct2012)
. format mydate %td
. list
```

```
        |    mydate |
        |-----------|
   1.   | 08oct2012 |
```

Now change the format to monthly:

```
. format mydate %tm
. list
```

|   | mydate |
|---|--------|
| 1. | 3566m3 |

You may think that the data have changed but to something absurd, so how did Stata mess up? In fact, the data have not changed. The data are still an integer value of 19274, which is October 8, 2012, when you count days from a zero date of January 1, 1960. The integer value 19274 is also March 3566 when you count months from a zero date of January 1960. The result is not what we may have wanted or expected, but Stata's point of view is that it gave us what we asked for, the same data value but interpreted according to a different format.

To convert a date, we need to use an appropriate conversion function. Then, and only then, will **format** work to produce nicer displays:

```
. generate mymonth = mofd(mydate)
. format mymonth %tm
. format mydate %td
. list
```

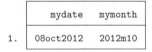

|   | mydate | mymonth |
|---|--------|---------|
| 1. | 08oct2012 | 2012m10 |

# Reference

Murphy, E. A. 1997. *The Logic of Medicine*. 2nd ed. Baltimore: Johns Hopkins University Press.

The Stata Journal (2013)
**13**, Number 1, pp. 217–219

# Stata tip 114: Expand paired dates to pairs of dates

Nicholas J. Cox
Department of Geography
Durham University
Durham, UK
n.j.cox@durham.ac.uk

Often dates for transactions or events come in pairs, for example, start and finish or open and close. People arrive for an appointment at a clinic and later leave; people order goods, which are later delivered; people start work for an employer and later leave. In all cases, when we take a snapshot, we may find people still in the clinic, goods not yet delivered, or (it is hoped) people still working for the employer.

With such events, it is natural that each pair of events is often recorded as an observation (case, row, or record) in a dataset. Such a structure makes some calculations easy. For example, the differences between times of arrivals and departures or of orders and deliveries are key to system performance, although the ideal will be short delays for seeing patients or selling goods and long delays for periods of employment or lifetimes. In Stata, with two variables named, say, `arrival` and `depart`, the delay is just

```
. generate lapse = depart - arrival
```

Precisely how we record time will depend on the problem, but here I am imagining a date or date–time or time variable.

If the closing events have yet to happen, then `depart` may need to be recorded as missing. If so, `lapse` will in turn be missing. Notice, by the way, a simple data quality check that the time lapse can never be negative. Time lapses recorded as zeros might also need checking in some situations: Was the system really that fast and efficient (or ruthless)?

Such a simple data structure—one observation for each time interval—may also be awkward, which leads to the main reason for this tip. Experience with data structures in Stata might lead readers to suggest a `reshape long`, which could be a good idea, but there is an easier alternative, to use `expand`.

We need first a unique or distinct identifier for each interval, which may already exist. The command `isid` allows a simple check of whether an identifier variable indeed matches its purpose. If the identifier variable is broken or nonexistent, then something like

```
. generate long id = _n
```

creates a new identifier fit for the purpose. Specifying a `long` variable type allows over two billion distinct positive identifiers, should they be needed. Otherwise, we use the existing identifier. Then

```
. expand 2
```

is the command needed. We turn each observation into two. The new observations are added at the end of the dataset, so we need to sort them before we can create two new variables that are the keys to other calculations. The first new variable combines the time information:

```
. by id, sort: generate time = cond(_n == 1, arrival, depart)
```

By virtue of the `expand` command, each distinct value of the identifier now occurs precisely twice. We can therefore use the framework provided by the `by:` prefix; see Cox (2002) for a tutorial if desired. Under `by:`, the observation number _n is interpreted within groups (here all pairs), and we assign time `arrival` to the first observation of two and time `depart` to the second.

Let's imagine a small section of a toy dataset and apply our expansion method.

```
. list
```

|     | id | arrival | depart |
|-----|----|---------|--------|
| 1.  | 1  | 1000    | 1100   |
| 2.  | 2  | 1100    | 1300   |
| 3.  | 3  | 1200    | 1400   |

```
. expand 2
(3 observations created)
. by id, sort: generate time = cond(_n == 1, arrival, depart)
```

The second new variable flags whether each time is an arrival or departure. The usual kind of indicator or dummy variable with values 1 or 0 would serve, but values of 1 and $-1$ are even better, given that each arrival is an addition and each departure a subtraction.

```
. by id: generate inout = cond(_n == 1, 1, -1)
```

Typing (_n == 1) - (_n == 2) is another way to code this.

Now all we need to do is sort and calculate the net results of all events:

```
. sort time
. list, separator(0)
```

|     | id | arrival | depart | time | inout |
|-----|----|---------|--------|------|-------|
| 1.  | 1  | 1000    | 1100   | 1000 | 1     |
| 2.  | 1  | 1000    | 1100   | 1100 | -1    |
| 3.  | 2  | 1100    | 1300   | 1100 | 1     |
| 4.  | 3  | 1200    | 1400   | 1200 | 1     |
| 5.  | 2  | 1100    | 1300   | 1300 | -1    |
| 6.  | 3  | 1200    | 1400   | 1400 | -1    |

Because the flag variable `inout` records additions and subtractions, so also its cumulative or running sum keeps track of the number inside the system. In a jargon common in economics, flows are used to calculate stocks, the assumption being that any stock from before the start of records would need to be added.

```
. generate present = sum(inout)
. list, separator(0)
```

|     | id | arrival | depart | time | inout | present |
|-----|----|---------|--------|------|-------|---------|
| 1.  | 1  | 1000    | 1100   | 1000 | 1     | 1       |
| 2.  | 1  | 1000    | 1100   | 1100 | -1    | 0       |
| 3.  | 2  | 1100    | 1300   | 1100 | 1     | 1       |
| 4.  | 3  | 1200    | 1400   | 1200 | 1     | 2       |
| 5.  | 2  | 1100    | 1300   | 1300 | -1    | 1       |
| 6.  | 3  | 1200    | 1400   | 1400 | -1    | 0       |

This is only one trick, and others will depend on your problem. For example, if a clinic is only open daily, the number present should drop to zero at the end of each day. More generally, stocks cannot be negative. The logic of how your system operates provides a logic for your code and checks on data quality. If that logic implies separate accounting for each panel of panel or longitudinal data, then that merely implies a different `sort` order and operations under the aegis of `by:` (Cox 2002).

For a different problem arising with paired data, see Cox (2008).

# References

Cox, N. J. 2002. Speaking Stata: How to move step by: step. *Stata Journal* 2: 86–102.

———. 2008. Stata tip 71: The problem of split identity, or how to group dyads. *Stata Journal* 8: 588–591.

The Stata Journal (2013)
**13**, Number 2, pp. 401–405

# Stata tip 115: How to properly estimate the multinomial probit model with heteroskedastic errors

Michael Herrmann
Department of Politics and Public Administration
University of Konstanz
Konstanz, Germany
michael.herrmann@uni-konstanz.de

Models for multinomial outcomes are frequently used to analyze individual decision making in consumer research, labor market research, voting, and other areas. The multinomial probit model provides a flexible approach to analyzing decisions in these fields because it does not impose some of the restrictive assumptions inherent in the often used conditional logit approach. In particular, multinomial probit relaxes 1) the assumption of independent error terms, allowing for correlation in individual choices across alternatives, and 2) it does not impose the assumption of identically distributed errors, allowing unobserved factors to affect the choice of some alternatives more strongly than others (that is, heteroskedasticity).

By default, `asmprobit` relaxes both the assumptions of independence and homoskedasticity. To avoid overfitting, however, the researcher may sometimes wish to relax these assumptions one at a time.[1] A seemingly straightforward solution would be to rely on the options `stddev()` and `correlation()`, which allow the user to set the structure for the error variances and their covariances, respectively (see [R] **asmprobit**).

When doing so, however, the user should be aware that specifying `std(het)` and `corr(ind)` does not actually fit a pure heteroskedastic multinomial probit model. With $J$ outcome categories, if errors are independent, $J - 1$ error variances are identified (see below). Instead, Stata estimates $J - 2$ error variances and, hence, imposes an additional constraint, which causes the model to be overidentified. As a result, the estimated model is not invariant to the choice of base and scale outcomes; that is, changing the base or scale outcome leads to different values of the likelihood function.

To properly estimate a pure heteroskedastic model, the user needs to define the structure of the error variances manually. This is easy to accomplish using the `pattern` or `fixed` option. The following example illustrates the problem and shows how to estimate the model correctly.

---

1. Another reason to relax them one at a time is that heteroskedasticity and error correlation cannot be distinguished from each other in the default specification. That is, one cannot simply look at the estimated covariance matrix of the errors and see whether the errors are heteroskedastic, correlated, or both. What Stata estimates is the normalized covariance matrix of error differences whose elements do not allow one to draw any conclusions on the covariance structure of the errors themselves.

Consider an individual's choice of travel mode with the alternatives being air, train, bus, and car and predictor variables, including general cost of travel, terminal time, household income, and traveling group size. One might suspect the choice of some alternatives to be driven more by unobserved factors than the choice of others. For example, there might be more unobserved reasons related to an individual's decision to travel by plane than by train, bus, or car. Allowing the error variances associated with the alternatives to differ, we fit the following model:

```
. use http://www.stata-press.com/data/r12/travel

. asmprobit choice travelcost termtime, casevars(income partysize)
> case(id) alternatives(mode) std(het) corr(ind) nolog
```

| Alternative-specific multinomial probit | | | | Number of obs | = | 840 |
| Case variable: id | | | | Number of cases | = | 210 |
| Alternative variable: mode | | | | Alts per case: min = | | 4 |
| | | | | avg = | | 4.0 |
| | | | | max = | | 4 |

```
Integration sequence:        Hammersley
Integration points:               200          Wald chi2(8)    =      71.57
Log simulated-likelihood = -181.81521          Prob > chi2     =     0.0000
```

| choice | Coef. | Std. Err. | z | P>\|z\| | [95% Conf. Interval] | |
|---|---|---|---|---|---|---|
| **mode** | | | | | | |
| travelcost | −.012028 | .0030838 | −3.90 | 0.000 | −.0180723 | −.0059838 |
| termtime | −.050713 | .0071117 | −7.13 | 0.000 | −.0646517 | −.0367743 |
| **air** | (base alternative) | | | | | |
| **train** | | | | | | |
| income | −.03859 | .0093287 | −4.14 | 0.000 | −.0568739 | −.0203062 |
| partysize | .7590228 | .190438 | 3.99 | 0.000 | .3857711 | 1.132274 |
| _cons | −.9960951 | .4750053 | −2.10 | 0.036 | −1.927088 | −.0651019 |
| **bus** | | | | | | |
| income | −.0119789 | .0081057 | −1.48 | 0.139 | −.0278658 | .003908 |
| partysize | .5876645 | .1751734 | 3.35 | 0.001 | .2443309 | .930998 |
| _cons | −1.629348 | .4803384 | −3.39 | 0.001 | −2.570794 | −.6879016 |
| **car** | | | | | | |
| income | −.004147 | .0078971 | −0.53 | 0.599 | −.019625 | .011331 |
| partysize | .5737318 | .163719 | 3.50 | 0.000 | .2528485 | .8946151 |
| _cons | −3.903084 | .750675 | −5.20 | 0.000 | −5.37438 | −2.431788 |
| /lnsigmaP1 | −1.097572 | .7967201 | −1.38 | 0.168 | −2.659115 | .4639704 |
| /lnsigmaP2 | −.3906271 | .3468426 | −1.13 | 0.260 | −1.070426 | .2891719 |
| sigma1 | 1 | (base alternative) | | | | |
| sigma2 | 1 | (scale alternative) | | | | |
| sigma3 | .3336802 | .2658497 | | | .0700102 | 1.590376 |
| sigma4 | .6766324 | .2346849 | | | .3428624 | 1.335321 |

```
(mode=air is the alternative normalizing location)
(mode=train is the alternative normalizing scale)
```

As can be seen, two of the four error variances are set to one. These are the base and scale alternatives. While choosing a base and scale alternative is necessary to identify the model, the problem here is that because errors are uncorrelated, fixing the variance of the base alternative is not necessary to identify the model. As a result, an additional constraint is imposed, which leads to a different model structure depending on the choice of base and scale alternatives. For example, changing the base alternative to car produces a different log likelihood:

```
. quietly asmprobit choice travelcost termtime, casevars(income partysize)
> case(id) alternatives(mode) std(het) corr(ind) nolog base(4)
. display e(ll)
-181.58795
```

To properly estimate an unconstrained heteroskedastic model, one needs to define a vector of variance terms in which one element (the scale alternative) is fixed and pass this vector on to the estimation command. For example, to set the error variance of the second alternative to unity, define a vector of missing values, stdpat, whose second element is 1, and then call this vector from inside asmprobit using the option std(fixed) (see [R] **asmprobit** for details):

```
. matrix define stdpat = (.,1,.,.)

. asmprobit choice travelcost termtime, casevars(income partysize)
> case(id) alternatives(mode) std(fixed stdpat) corr(ind) nolog base(1)
```

| Alternative-specific multinomial probit | | Number of obs | = | 840 |
| Case variable: id | | Number of cases | = | 210 |

| Alternative variable: mode | | Alts per case: min = | 4 |
| | | avg = | 4.0 |
| | | max = | 4 |

| Integration sequence: | Hammersley | | | |
| Integration points: | 200 | Wald chi2(8) | = | 26.84 |
| Log simulated-likelihood = -180.01839 | | Prob > chi2 | = | 0.0008 |

| choice | Coef. | Std. Err. | z | P>\|z\| | [95% Conf. Interval] | |
|---|---|---|---|---|---|---|
| mode | | | | | | |
| travelcost | -.0196389 | .0067143 | -2.92 | 0.003 | -.0327988 | -.006479 |
| termtime | -.0664153 | .0140353 | -4.73 | 0.000 | -.093924 | -.0389065 |
| air | (base alternative) | | | | | |
| train | | | | | | |
| income | -.0498732 | .0154884 | -3.22 | 0.001 | -.08023 | -.0195165 |
| partysize | 1.126922 | .3651321 | 3.09 | 0.002 | .4112761 | 1.842568 |
| _cons | -1.072849 | .680711 | -1.58 | 0.115 | -2.407018 | .2613198 |
| bus | | | | | | |
| income | -.0210642 | .0139892 | -1.51 | 0.132 | -.0484826 | .0063542 |
| partysize | .8678651 | .3179559 | 2.73 | 0.006 | .244683 | 1.491047 |
| _cons | -1.831363 | .7345686 | -2.49 | 0.013 | -3.271091 | -.3916349 |
| car | | | | | | |
| income | -.010205 | .0131711 | -0.77 | 0.438 | -.0360199 | .01561 |
| partysize | .8708577 | .3202671 | 2.72 | 0.007 | .2431458 | 1.49857 |
| _cons | -4.971594 | 1.261002 | -3.94 | 0.000 | -7.443112 | -2.500075 |
| /lnsigmaP1 | .558377 | .3076004 | 1.82 | 0.069 | -.0445087 | 1.161263 |
| /lnsigmaP2 | -1.0078 | 1.116358 | -0.90 | 0.367 | -3.195822 | 1.180223 |
| /lnsigmaP3 | -.0158072 | .3593511 | -0.04 | 0.965 | -.7201225 | .6885081 |
| sigma1 | 1.747833 | .5376342 | | | .9564673 | 3.193964 |
| sigma2 | 1 | (scale alternative) | | | | |
| sigma3 | .3650213 | .4074946 | | | .0409329 | 3.255099 |
| sigma4 | .9843171 | .3537155 | | | .4866926 | 1.990743 |

```
(mode=air is the alternative normalizing location)
(mode=train is the alternative normalizing scale)
```

Now the model is properly normalized, and the user may verify that changing either the scale alternative (that is, changing the location of the 1 in `stdpat`) or the base alternative leaves results unchanged. Note that while, in theory, the only restriction necessary to identify the heteroskedastic probit model is to fix one of the variance terms, in the Stata implementation of the model, the base and scale outcomes must be different. That is, Stata does not allow the same alternative to be the base outcome and the scale outcome. However, this is more of an inconvenience than a restriction: such a model would be equivalent to one in which the base and scale outcomes differed.

Finally, to show that independence of errors indeed implies $J - 1$ estimable error variances, we must verify that the error variances can be calculated directly from the variance and covariance parameters of the normalized error differences. Only the latter are identified and, hence, estimable (Train 2009). Suppose, without loss of generality, $J = 3$, and let $j = 1$ be the base outcome.

Following the normalization approach advocated by Train (2009, 100f.), the normalized covariance matrix of error differences is given by

$$\widetilde{\Omega}_1^* = \begin{pmatrix} 1 & \theta_{23}^* \\ & \theta_{33}^* \end{pmatrix}$$

with elements $\theta^*$ relating to the actual error variances $\sigma_{jj}$ and covariances $\sigma_{ij}$ as follows:

$$\theta_{23}^* = \frac{\sigma_{23} + \sigma_{11} - \sigma_{12} - \sigma_{13}}{\sigma_{22} + \sigma_{11} - 2\sigma_{12}}$$

$$\theta_{33}^* = \frac{\sigma_{33} + \sigma_{11} - 2\sigma_{13}}{\sigma_{22} + \sigma_{11} - 2\sigma_{12}}$$

Under independence, $\sigma_{ij} = 0$. Fixing $\sigma_{22} = 1$ (that is, choosing $j = 2$ as the scale outcome) yields $\theta_{23}^* = \sigma_{11}/(1 + \sigma_{11})$ and $\theta_{33}^* = (\sigma_{33} + \sigma_{11})/(1 + \sigma_{11})$. Obviously, $\sigma_{11}$ can be calculated from $\theta_{23}^*$, and subsequent substitution produces $\sigma_{33}$ from $\theta_{33}^*$. The same is true if we choose to fix either $\sigma_{11}$ or $\sigma_{33}$ because in each case, we would obtain two equations in two unknowns. Similar conclusions follow when there are four or more outcome categories. Thus, with independent errors, $J - 1$ variance parameters are estimable.

# Reference

Train, K. E. 2009. *Discrete Choice Methods with Simulation*. 2nd ed. Cambridge: Cambridge University Press.

The Stata Journal (2014)
**14**, Number 1, pp. 218–220

# Stata tip 116: Where did my p-values go? (Part 3)

Maarten L. Buis
Wissenschaftszentrum Berlin für Sozialforschung (WZB)
Berlin, Germany
maarten.buis@wzb.eu

In a previous Stata tip (Buis 2007), I discussed how to recover $t$ statistics, $p$-values, and confidence intervals for regression parameters by using the results that are returned by an estimation command. In a subsequent Stata tip (Buis 2011), I discussed how to recover parameter estimates for parameters that were estimated on a transformed scale. For example, if a likelihood function contains a standard deviation or a correlation, then many Stata commands will maximize the likelihood with respect to ln(standard deviation) and the Fisher's $z$ transformation of the correlation. In this tip, I will discuss how to recover the standard errors for the back-transformed parameters, that is, the standard errors of the standard deviation and the correlation.

Stata often displays the back-transformed parameters and their standard errors, but it leaves behind only the estimates of the transformed parameters and their standard errors. In those cases, the delta method (for example, Feiveson 2005) was used to compute the standard errors of the back-transformed parameters. In its simplest form, the delta method means that if we apply a transformation $G(\cdot)$ to a parameter $b$, then we can approximate the standard error of the transformed parameter as

$$\mathrm{se}\{G(b)\} \approx \mathrm{se}(b) \times G'\left(\widehat{b}\right)$$

where $G'(\widehat{b})$ is the first derivative of $G(b)$ with respect to $b$ evaluated at $\widehat{b}$. If Stata returned ln(standard deviation) and we wanted the standard deviation and its standard error, then $G(b) = \exp(b)$ and $G'(\widehat{b}) = \exp(\widehat{b})$. If Stata returned Fisher's $z$ transformation of a correlation and we wanted the correlation and its standard error, then $G(b) = \tanh(b)$ and $G'(\widehat{b}) = \cosh(\widehat{b})^{-2}$. This is illustrated below using a model estimated with `heckman` (see [R] **heckman**). This model was chosen because it returns transformed parameters of both types.

```
. use http://www.stata-press.com/data/r13/womenwk

. heckman wage educ, select(married children educ) nolog
Heckman selection model                       Number of obs    =        2000
(regression model with sample selection)      Censored obs     =         657
                                              Uncensored obs   =        1343

                                              Wald chi2(1)     =      403.39
Log likelihood = -5250.348                    Prob > chi2      =      0.0000
```

| wage | Coef. | Std. Err. | z | P>\|z\| | [95% Conf. Interval] | |
|---|---|---|---|---|---|---|
| **wage** | | | | | | |
| education | 1.099506 | .0547435 | 20.08 | 0.000 | .9922102 | 1.206801 |
| _cons | 7.042147 | .8423253 | 8.36 | 0.000 | 5.39122 | 8.693074 |
| **select** | | | | | | |
| married | .5420304 | .0657798 | 8.24 | 0.000 | .4131044 | .6709564 |
| children | .4409418 | .0276093 | 15.97 | 0.000 | .3868286 | .495055 |
| education | .0722993 | .0105096 | 6.88 | 0.000 | .0517007 | .0928978 |
| _cons | -1.473038 | .1465476 | -10.05 | 0.000 | -1.760266 | -1.18581 |
| /athrho | .8081049 | .1108545 | 7.29 | 0.000 | .5908341 | 1.025376 |
| /lnsigma | 1.807547 | .0291035 | 62.11 | 0.000 | 1.750506 | 1.864589 |
| rho | .6685435 | .061308 | | | .5304953 | .772047 |
| sigma | 6.095479 | .1773995 | | | 5.757513 | 6.453283 |
| lambda | 4.075093 | .4690025 | | | 3.155865 | 4.994321 |

```
LR test of indep. eqns. (rho = 0):   chi2(1) =      47.02   Prob > chi2 = 0.0000

. tempname gprime

. scalar `gprime' = exp(_b[lnsigma:_cons])

. display "se of sigma = " _se[lnsigma:_cons]*`gprime'
se of sigma = .17739954

. scalar `gprime' = cosh(_b[athrho:_cons])^-2

. display "se of rho = " _se[athrho:_cons]*`gprime'
se of rho = .06130802
```

Alternatively, one can use nlcom (see [R] **nlcom**) to compute these standard errors,
as follows:

```
. nlcom (rho: tanh(_b[athrho:_cons])) (sigma: exp(_b[lnsigma:_cons]))
        rho:  tanh( _b[athrho:_cons]  )
      sigma:  exp( _b[lnsigma:_cons]  )
```

| wage | Coef. | Std. Err. | z | P>\|z\| | [95% Conf. Interval] | |
|---|---|---|---|---|---|---|
| rho | .6685435 | .061308 | 10.90 | 0.000 | .548382 | .788705 |
| sigma | 6.095479 | .1773995 | 34.36 | 0.000 | 5.747782 | 6.443175 |

Notice that the confidence intervals do not correspond with those in the output of
heckman. This is because heckman first computes the bounds of the confidence intervals
for the transformed parameters and then back-transforms those bounds to the original
metric, while nlcom uses the standard errors for the back-transformed parameters for
computing these bounds. In most cases, computing the bounds on the transformed scale

and then back-transforming those bounds to the original scale results in somewhat better bounds. This is because the sampling distribution of the transformed parameters is likely to be better approximated by a normal distribution than is the sampling distribution of the back-transformed parameters. For more discussion, see Sribney and Wiggins (2009). You can use the techniques discussed in Buis (2007) to recover the confidence intervals reported by `heckman`.

```
. display "confidence interval for rho: ["
> tanh( _b[athrho:_cons] - invnormal(.975)*_se[athrho:_cons] ) ", "
> tanh( _b[athrho:_cons] + invnormal(.975)*_se[athrho:_cons] ) "]"
confidence interval for rho: [.53049526, .77204703]
. display "confidence interval for sigma: ["
> exp( _b[lnsigma:_cons] - invnormal(.975)*_se[lnsigma:_cons] ) ", "
> exp( _b[lnsigma:_cons] + invnormal(.975)*_se[lnsigma:_cons] ) "]"
confidence interval for sigma: [5.7575126, 6.4532831]
```

Also note that `nlcom` returns the $z$ statistic and $p$-value for the test of the null hypothesis that the standard deviation and the correlation are 0, which were not reported by `heckman`. This test is problematic in the case of the standard deviation, because this is a test "on the boundary of the parameter space". A standard deviation can only take values larger than or equal to 0. Thus the hypothesis that the standard deviation is equal to 0 is on the boundary of the possible values for the standard deviation, and standard tests do not tend to behave well in this extreme area (for example, Gutierrez, Carter, and Drukker [2001]).

# References

Buis, M. L. 2007. Stata tip 53: Where did my p-values go? *Stata Journal* 7: 584–586.

———. 2011. Stata tip 97: Getting at $\rho$'s and $\sigma$'s. *Stata Journal* 11: 315–317.

Feiveson, A. H. 2005. FAQ: What is the delta method and how is it used to estimate the standard error of a transformed parameter?
http://www.stata.com/support/faqs/statistics/delta-method/.

Gutierrez, R. G., S. Carter, and D. M. Drukker. 2001. sg160: On boundary-value likelihood-ratio tests. *Stata Technical Bulletin* 60: 15–18. Reprinted in *Stata Technical Bulletin Reprints*, vol. 10, pp. 269–273. College Station, TX: Stata Press.

Sribney, W., and V. Wiggins. 2009. FAQ: Standard errors, confidence intervals, and significance tests for ORs, HRs, IRRs, and RRRs.
http://www.stata.com/support/faqs/statistics/delta-rule/.

The Stata Journal (2014)
**14**, Number 1, pp. 221–225

# Stata tip 117: graph combine—Combining graphs

Lars Ängquist
Institute of Preventive Medicine
Bispebjerg and Frederiksberg Hospitals—The Capital Region
Copenhagen, Denmark
lars.henrik.angquist@regionh.dk

## 1 Introduction

There are many different reasons for wanting to create multipanel graphs, presented in $r \geq 1$ rows and $c \geq 1$ columns: these reasons include making efficient use of restricted display space and enhancing the presentation of results. In basic Stata, the flexible approach to confidently handle these tasks is given by using the `graph combine` functionality (see `help graph combine`). For related discussions and examples, see the stimulating books *An Introduction to Stata for Health Researchers* (Juul and Frydenberg 2010) and *A Visual Guide to Stata Graphics* (Mitchell 2012).

## 2 Basic usage

First, we start with setting up seven simple, but quite artificial, linear relations disturbed by normally distributed noise based on simulated $x$ and $y$ variables (interpreted in the standard sense).

```
set obs 100
generate xvar=10*runiform()
forvalues i=1/7 {
    generate y`i'=xvar*`i'+runiform()*(`i'*3)
    label variable y`i' "Outcome variable `i'"
}
```

Second, we simply fit linear regressions that correspond to these relations and then save the seven corresponding graphs in memory.

```
foreach yvar of varlist y* {
    local lbl: variable label `yvar'
    sort xvar
    reg `yvar' xvar
    local b : display %3.2f _b[xvar]
    predict p, xb
    twoway (scatter `yvar' xvar) (line p xvar),        ///
        ytitle("`lbl'") xtitle("Explanatory covariate") ///
        yscale(range(0 80))                            ///
        legend(off) note("{&beta}=`b'", position(4) ring(0)) ///
        name("graph_`yvar'", replace)
    drop p
}
```

(Here we use the `name(`*string*`)` option—unless we want to actually save the separate graphs to disk. In that case, we would replace this option with `saving(`*string*`)`.)

Finally, we intend to combine the graphs into a multipanel setup. Assuming that the graphs belong to two distinct groups—graphs 1–3 and 4–7, respectively—they are mirrored in the construction. This is achieved by the following:

1. Combine graphs 1–3 into panel 1.

2. Combine graphs 4–7 into panel 2.

3. Combine the resulting 1-row panels, panel 1 ($r \times c = 1 \times 3$) and panel 2 ($r \times c = 1 \times 4$), into a final 2-row panel ($r = 2$; see figure 1).

```
graph combine graph_y1 graph_y2 graph_y3,                              ///
    name("firstset", replace) ycommon cols(3) title("First set of graphs")
graph combine graph_y4 graph_y5 graph_y6 graph_y7,                    ///
    name("secondset", replace) ycommon cols(4) title("Second set of graphs")

graph combine firstset secondset,                                     ///
    saving("sevenpanelgraph.gph", replace) ycommon cols(1)
graph export sevenpanelgraph.eps, replace
```

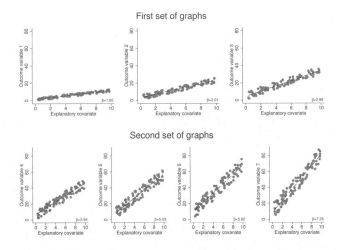

Figure 1. Multipanel graph—a combination of combined graphs

## 3   Some notes on options

The basic functionality facilitates an easy-to-use combination of graphs. A well-suited set of selected options might improve the display.

### 3.1 Axes

In many cases, keeping scales constant over panels might enhance the interpretability of the jointly graphed relations. Generally, this might prove to be a valid argument; however, it is imperative for the $x$ axis and $y$ axis when comparing vertically (the `xcommon` option) and horizontally (the `ycommon` option), respectively.

### 3.2 Margins

To keep the panels as tightly linked as possible—to increase overall comparability—it might be suitable to reduce margins through `imargin(zero)`; for other margin choices, see `help marginstyle`.

### 3.3 Panel pattern

The final number of panels to use is implicitly given by the stated list of panels in the actual program call. (Remember that each panel might in itself be a previously constructed multipanel. In the above example, a single column, $c = 1$, was used at the combination stage.) To define which $r \times c$ panel-matrix shape will be used, one may choose any of the following options (one is enough): `rows(integer)` or `cols(integer)`. To make the graph (distribution of panels) unique, select the `colfirst` option (or not). If the required number of panels is less than the available number $r{\cdot}c$, it may be useful to explicitly—given the unique order—tell Stata which panels should be left empty (instead of the default) by using `holes(numlist)`.

### 3.4 Scaling

Each panel is downscaled when using multipanels, text and markers, etc. It is possible to rescale the downscaling through the `iscale(scale)` option, where *scale* is either an absolute (positive) or a relative value. For example, the absolute value 1 means the original size, and the relative value `*1` implies the same size as the default selection; 0.75 and `*0.75` will adjust the size to the three-quarter size counterparts.

## 4    A second example

For our second example, we will play around with the individual panel sizes. For this, we will use one of the seven graphs (the sixth) from figure 1, which is inspired by the informative help file (see the end of the `help graph combine` post), to complement it with the two corresponding underlying histograms (see result in figure 2).

```
histogram xvar,                                    ///
    percent start(0) width(1)                      ///
    xscale(range(0 10) off)                        ///
    fxsize(100) fysize(25)                         ///
    yscale(range(0 15)) ytitle("")                 ///
    ylabel(0(5)15, angle(horizontal))              ///
    kdensity kdenopts(lpattern(dash))              ///
    plotregion(margin(zero))                       ///
    note("N (%)", ring(0) position(10))            ///
    name("hist_xvar", replace)
histogram y6,                                      ///
    percent start(0) width(10) horizontal          ///
    xtitle("") xlabel(0(10)20) xscale(rev)         ///
    fxsize(25) fysize(100)                         ///
    yscale(range(0 80) off)                        ///
    ylabel(10(20)70, angle(horizontal))            ///
    kdensity kdenopts(lpattern(dash))              ///
    plotregion(margin(zero))                       ///
    note("N (%)", ring(0) position(4))             ///
    name("hist_y6", replace)
```

In the next step, these three panels are combined (note that we use some of the options just discussed). The main point here is that the options fxsize(*number*) and fysize(*number*) govern the widths and heights of the panels; that is, in the example above, the thin sides are left at 25% of the original sizes.

```
graph combine hist_y6 graph_y6 hist_xvar,         ///
    holes(3) rows(2)                               ///
    imargin(0 2 0 0)                               ///
    title(" Twoway graph with histograms", ring(0))   ///
    saving(graphwithhistograms.gph, replace)
graph export graphwithhistograms.eps, replace
```

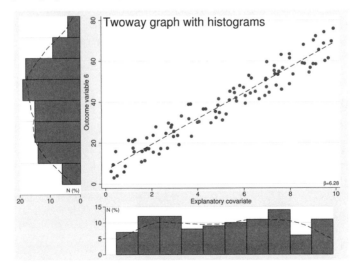

Figure 2. Multipanel graph—a scatterplot with a prediction line and two complementary histograms

# 5    Discussion and alternatives

In many situations where the subgraphs combine corresponding true data subsets of the present loaded data, a similarly performing alternative would be to use the by() option (see `help by_option`). Here the syntax by(*varlist*[ , *options* ]) allows combined graphing of the corresponding defined graph with respect to all present categories specified by the categorical variables given in *varlist*. In this setting, the options `total` and `missing` add panels based on the total dataset (over nonmissing groups) and missing data for individuals, respectively.

## 5.1    by() options

As noted above, the option by() allows for suboptions. Some suboptions are similar to the ones available for `graph combine`—for example, `colfirst`, `cols()`, `rows()`, `holes()`, `iscale()`, and `imargin()`. Similar functionality, but with different names and adapted settings, is given by `compact` (reduces margins between panels), `norescale` (uses the same scales over panels), and `noedgelabel` (restricts the number of displayed labels). Note that an option with `no`, such as `norescale`, generally has a counterpart, such as `rescale`, with a quite obvious implication.

Usually, this type of solution might be convenient in different cases; however, in most situations, this solution is less flexible and more restrictive by nature. Furthermore, graphing several subgroups within a single panel (together but separately marked) is an alternative solution that allows the smaller number of subgroups to be totally displayed while applying distinct colors and markers. For other cases, the multipanel design may be the best choice because one (or several) background groups can be added to each panel to enhance overall comparability. For example, see the discussion of overlaid graphs in Cox (2010), where subgroups are plotted against completely complementary data while using discrete gray-scaled backdrop markers for the background group. This is referred to as adopting a substrate, or subset, graphing design.

# References

Cox, N. J. 2010. Speaking Stata: Graphing subsets. *Stata Journal* 10: 670–681.

Juul, S., and M. Frydenberg. 2010. *An Introduction to Stata for Health Researchers.* 3rd ed. College Station, TX: Stata Press.

Mitchell, M. N. 2012. *A Visual Guide to Stata Graphics.* 3rd ed. College Station, TX: Stata Press.

The Stata Journal (2014)
**14**, Number 1, pp. 226–229

# Stata tip 118: Orthogonalizing powered and product terms using residual centering

Carsten Sauer
Bielefeld University
Bielefeld, Germany
carsten.sauer@uni-bielefeld.de

In multiple regressions, powered variables are commonly included to represent higher-order nonlinear effects. Interaction effects are often represented by the product of two variables. In many cases, including powered or product terms leads to increasing correlation or collinearity. This correlation between the powered or the product term and the first-order predictor variable can lead to unstable estimates and bouncing beta weights. To tackle this problem, one can center the predictor variables before the estimation at the means (Cohen 1978). An alternative approach to mean centering is residual centering. Unlike mean centering, residual centering ensures orthogonality between the powered or the product term and the first-order predictor variables. The theoretical background of residual centering in comparison with mean centering is described in Little, Bovaird, and Widaman (2006). Technically, residual centering is a two-stage ordinary least-squares regression procedure introduced by Lance (1988). In the first step, the powered term or the product term is regressed on the first-order predictor variables. The residuals of this regression are then used to represent the powered or the product term. The variance of the powered or the product term obtained by this regression is independent of the variance of the first-order predictor variable.

The following examples show the residual centering method for powered and product terms. First, we run a regression with one linear effect; then we include the powered term.

```
. sysuse auto
(1978 Automobile Data)
. quietly regress price weight
. estimates store M1
. generate weight2 = weight^2
. quietly regress price weight weight2
. estimates store M2
. estat vif
```

| Variable | VIF | 1/VIF |
|---|---|---|
| weight | 58.95 | 0.016963 |
| weight2 | 58.95 | 0.016963 |
| Mean VIF | 58.95 | |

The variance inflation factor is 58.95, which indicates multicollinearity. Now we calculate the residual-centered powered term and rerun the regression model.

```
. quietly regress weight2 weight
. predict weight2_rc, res
. quietly regress price weight weight2_rc
. estimates store M3
. estat vif
    Variable |       VIF       1/VIF
-------------+----------------------
      weight |      1.00    1.000000
  weight2_rc |      1.00    1.000000
-------------+----------------------
    Mean VIF |      1.00
```

The test for collinearity indicates that perfect orthogonality could be achieved via residual centering. The following table shows the coefficients of the models with only the linear effect (M1), the quadratic effects without residual centering (M2), and the quadratic effects with residual centering (M3). The results show that including the interaction term in the third model does not change the coefficients of the first-order variable and the constant term. Note that residual centering does not affect the proportion of explained variance ($R^2$).

```
. estimates table M1 M2 M3, b(%8.3f) stats(r2 N) star
```

| Variable | M1 | M2 | M3 |
|---|---|---|---|
| weight | 2.044*** | −7.273** | 2.044*** |
| weight2 | | 0.002*** | |
| weight2_rc | | | 0.002*** |
| _cons | −6.707 | 1.3e+04** | −6.707 |
| r2 | 0.290 | 0.394 | 0.394 |
| N | 74 | 74 | 74 |

legend: * p<0.05; ** p<0.01; *** p<0.001

The second example shows the residual centering for the interaction terms of two continuous variables, weight and length.

```
. quietly regress price weight length
. estimates store M4
. generate interact = weight*length
. quietly regress price weight length interact
. estimates store M5
. estat vif
    Variable |       VIF       1/VIF
-------------+----------------------
     interact|    155.59    0.006427
      weight |     91.62    0.010915
      length |     21.84    0.045795
-------------+----------------------
    Mean VIF |     89.68
```

```
. quietly regress interact weight length
. predict interact_rc, res
. quietly regress price weight length interact_rc
. estimates store M6
```

The test for collinearity indicates results similar to those shown above. The variance inflation factor is very high for the interaction model without centering and remarkably lower for the interaction model with residual centering.

```
. estat vif
    Variable |       VIF       1/VIF
-------------+----------------------
      length |      9.52    0.105068
      weight |      9.52    0.105068
 interact_rc |      1.00    1.000000
-------------+----------------------
    Mean VIF |      6.68
```

The following table shows the coefficients for the two first-order predictor variables (M4), the interaction effect without residual centering (M5), and the interaction effect with residual centering (M6). Again including the orthogonalized product term does not change the coefficients for the first-order predictor variables and the constant term.

```
. estimates table M4 M5 M6, b(%8.3f) stats(r2 N) star
```

| Variable | M4 | M5 | M6 |
|---|---|---|---|
| weight | 4.699*** | -0.967 | 4.699*** |
| length | -97.960* | -174.570** | -97.960* |
| interact | | 0.029 | |
| interact_rc | | | 0.029 |
| _cons | 1.0e+04* | 2.5e+04** | 1.0e+04* |
| r2 | 0.348 | 0.375 | 0.375 |
| N | 74 | 74 | 74 |

```
legend: * p<0.05; ** p<0.01; *** p<0.001
```

Residual centering is an easy two-step procedure that makes it possible to obtain unbiased powered or interaction effects and main effects. While it was originally used for multiple regression models, Little, Bovaird, and Widaman (2006) extend the residual centering procedure to represent latent variable interactions in structural equation modeling. Geldhof et al. (2012) provide an overview about applications and caveats for latent interactions.

# References

Cohen, J. 1978. Partialed products *are* interactions; partialed powers *are* curve components. *Psychological Bulletin* 85: 858–866.

Geldhof, G. J., S. Pornprasertmanit, A. M. Schoemann, and T. D. Little. 2012. Orthogonalizing through residual centering: Extended applications and caveats. *Educational and Psychological Measurement* 73: 27–46.

Lance, C. E. 1988. Residual centering, exploratory and confirmatory moderator analysis, and decomposition of effects in path models containing interactions. *Applied Psychological Measurement* 12: 163–175.

Little, T. D., J. A. Bovaird, and K. F. Widaman. 2006. On the merits of orthogonalizing powered and product terms: Implications for modeling interactions among latent variables. *Structural Equation Modeling: A Multidisciplinary Journal* 13: 497–519.

The Stata Journal (2014)
**14**, Number 1, pp. 230–235

# Stata tip 119: Expanding datasets for graphical ends

Nicholas J. Cox
Department of Geography
Durham University
Durham, UK
n.j.cox@durham.ac.uk

Graphical tasks in Stata range from direct plotting of the data in memory (scatterplots, for example) to plotting of results calculated on the fly (histograms, for example). With the first kind of problem, the only issues are graphical. With the second kind of problem, Stata commands need to do some work on your behalf before graphs can be drawn. For a histogram, Stata needs to calculate bar coordinates (heights and base locations) before they can be plotted. Such preliminary work is often needed, although much of the art of graphics programming is to hide it from the user.

When Stata graphical commands are not available for what you want to do, changing the dataset temporarily is often the strategy to adopt. Here I show how using the expand (see [D] **expand**) command is a way to overcome some otherwise awkward challenges.

With the familiar auto dataset, we could look at means for, say, mpg (miles per gallon) for a one-way breakdown of the data and for a two-way breakdown of the data. Here we classify by the categorical variables foreign (whether cars are domestic or foreign, meaning U.S. made) and rep78 (repair record).

```
. sysuse auto
(1978 Automobile Data)
. graph dot (mean) mpg, over(foreign) ytitle(Mean miles per gallon) nofill
> missing
. graph dot (mean) mpg, over(rep78) over(foreign) ytitle(Mean miles per gallon)
> nofill missing
```

A detail here is specifying the missing option. Clearly, if values missing on one or another categorical variable were of no concern to us, we would not do that. Either way, being careful about missing values is a good idea. Figure 1 shows the two graphs side by side.

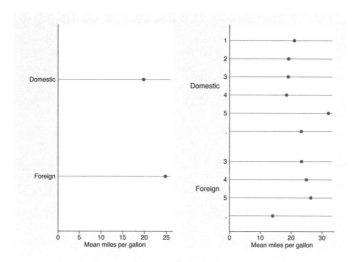

Figure 1. Two displays of mean miles per gallon. The left-hand graph is a poor use of space, but including its means in the right-hand graph would make sense.

Few researchers might want to show the left-hand graph of figure 1 because it conveys too little information. But many might want to enhance the right-hand graph with the summaries it shows. So how do we do that? The two graphs show different reductions of the data, so we need to ensure that both reductions are accessible at the same time. After saving a copy of the dataset first, we can double the dataset with `expand`:

```
. preserve
. expand 2
(74 observations created)
```

`expand` here adds an extra copy of the data as extra observations so that the first half of the observations is the original dataset and the second half is new but identical. Knowing that `rep78` has values 1 to 5 and that there are some missings (.), we could lump all the values together using any other integer or extended missing value.

```
. replace rep78 = .z if _n > _N/2
(74 real changes made, 74 to missing)
. label define rep78 .z "all"
. label values rep78 rep78
. graph dot (mean) mpg, over(rep78) over(foreign) ytitle(Mean miles per gallon)
> nofill missing
```

Figure 2 shows the result. In effect, we did a two-way breakdown with one half of the dataset and a one-way breakdown with the other half, but given the way we structured the data, Stata ended up doing both at once.

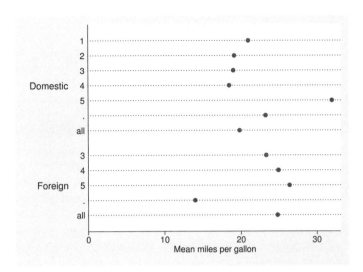

Figure 2. Extended summary of mean miles per gallon combining one-way and two-way breakdowns.

The odd-looking choice of `.z` to show results for "all" deserves comment. You can attach a value label to it, as just shown, so it will not be plotted as such. One particular reason for choosing `.z` is that it is larger than anything else, even when system missings (`.`) are present. Hence, it will be plotted at the end of each block of lines in the graph. Similarly, if you were using **graph bar** or **graph hbar** to draw bar charts, it would be plotted at the end of each block of bars. Either way, the **missing** option is now essential. It is possible that you are already using `.z` as a value. In that case, or for other reasons, you could instead choose any integer that ensures results for "all" are plotted at the beginning of each block of lines or bars. In this example, 0 or any negative integer would work fine.

Figure 2 is only a start. We might want to improve it in some way—say, by using different marker symbols for the "all" category—but we will leave the example there.

The same dataset can be used to show another application of **expand** to solve a similar graphical challenge. Imagine starting again and drawing some box plots. This time, we choose to be indifferent to the missing values.

```
. sysuse auto, clear
(1978 Automobile Data)
. graph box mpg, over(rep78)
. graph box mpg, over(foreign)
```

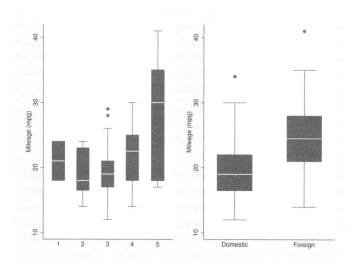

Figure 3. Box plots for miles per gallon classified by repair record (left) and domestic or foreign (right). The two graphs were produced separately and put together with `graph combine`.

We could just juxtapose these box plots, as in figure 3, but now there are two related small problems. The different sizes (bar widths) of the boxes are disconcerting, and the display for repair record takes up too much space for the information given. We could tinker with the sizes of the graphs, but we will explore a solution based on an initial `expand`.

We `expand` the data and produce a combined categorical variable. We cannot just copy `foreign`: its values of 0 and 1 overlap with those of 1 to 5 for `rep78`. Adding 6 to `foreign` will circumvent this problem.

```
. preserve
. expand 2
(74 observations created)
. generate x = cond(_n < _N/2, rep78, 6 + foreign)
(5 missing values generated)
. label define x 6 "Domestic" 7 "Foreign"
. label values x x
```

We now want something like `graph box mpg, over(x)`. With some more work, we can go further. `graph` will use different colors, markers, and so forth if it is plotting different variables. The `separate` command is ideal for producing two or more response variables (graphically, *y* variables) from one.

```
. separate mpg, by(x > 5)

                 storage   display   value
variable name    type      format    label      variable label

mpg0             byte      %8.0g                 mpg, !(x > 5)
mpg1             byte      %8.0g                 mpg, x > 5
. graph box mpg?, over(x, relabel(3 `" "3" "Repair record" "´)) nofill
> legend(off) box(1, fcolor(white)) box(2, fcolor(gs4))
> medline(lc(gs8)) ytitle("`: var label mpg´") ylabel(, angle(h))
```

Figure 4 shows the result. Small details include using `relabel()` to provide a label for repair record, centered appropriately; tinkering with default colors for the box display; and getting an axis title directly from the variable label of `mpg`. (Start with `macro` (see [P] **macro**) if you want more information on the syntax used for the last detail.)

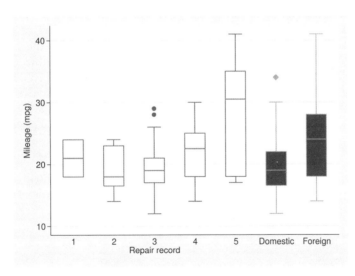

Figure 4. Box plots for miles per gallon classified by repair record (left) and domestic or foreign (right). The two graphs were produced in one call after rearranging the data.

This tip aims at giving you just a taste of what can be done with `expand` to solve some graphical problems, but more can be said.

Recall that each example using **expand** first used **preserve**, saving whatever data were in memory. If you did this within a do-file or program, then by default, there would be an automatic **restore**. With a **saved** copy of the dataset, you can afford to be cavalier about changing the data.

Some graphical problems might require **expand 3** or even more copies of the data. If that is so, know that **expand** adds observations to the existing dataset observation by observation rather than in blocks. Thus with **expand 3**, the original dataset would be followed by two copies of the first observation, two copies of the second observation, and so forth. If you have an identifier for each observation, such as **make** in the auto dataset, an identifier distinguishing different copies of the original would be useful, as in

```
. by make, sort: generate block = _n
```

If you did not have such an identifier, you would need to create one before the **expand**, by typing

```
. generate long id = _n
```

All of this is naturally based on the assumption that memory is available to allow the **expand** command to run. You could ease any memory problem by ensuring that variables or observations not needed for graphics were dropped with the **drop** command first. If that is insufficient, you might be able to use similar ideas but ones based on adding extra observations containing appropriate summaries. That could be very easy or very difficult. In the first example here, adding two observations with the means of **mpg** for the two categories of **foreign** would be enough. In the second example, adding enough information to produce the same box plots would be much more challenging.